NATO ASI Series
Advanced Science Institutes Series

A series presenting the results of activities sponsored by the NATO Science Committee, which aims at the dissemination of advanced scientific and technological knowledge, with a view to strengthening links between scientific communities.

The Series is published by an international board of publishers in conjunction with the NATO Scientific Affairs Division.

A Life Sciences	Plenum Publishing Corporation
B Physics	London and New York
C Mathematical and Physical Sciences	Kluwer Academic Publishers
D Behavioural and Social Sciences	Dordrecht, Boston and London
E Applied Sciences	
F Computer and Systems Sciences	Springer-Verlag
G Ecological Sciences	Berlin Heidelberg New York Barcelona
H Cell Biology	Budapest Hong Kong London Milan
I Global Environmental Change	Paris Santa Clara Singapore Tokyo

Partnership Sub-Series

1. Disarmament Technologies	Kluwer Academic Publishers
2. Environment	Springer-Verlag
3. High Technology	Kluwer Academic Publishers
4. Science and Technology Policy	Kluwer Academic Publishers
5. Computer Networking	Kluwer Academic Publishers

The Partnership Sub-Series incorporates activities undertaken in collaboration with NATO's Cooperation Partners, the countries of the CIS and Central and Eastern Europe, in Priority Areas of concern to those countries.

NATO-PCO Database

The electronic index to the NATO ASI Series provides full bibliographical references (with keywords and/or abstracts) to about 50 000 contributions from international scientists published in all sections of the NATO ASI Series. Access to the NATO-PCO Database compiled by the NATO Publication Coordination Office is possible in two ways:

– via online FILE 128 (NATO-PCO DATABASE) hosted by ESRIN, Via Galileo Galilei, I-00044 Frascati, Italy.

– via CD-ROM "NATO Science & Technology Disk" with user-friendly retrieval software in English, French and German (© WTV GmbH and DATAWARE Technologies Inc. 1992).

The CD-ROM can be ordered through any member of the Board of Publishers or through NATO-PCO, B-3090 Overijse, Belgium.

Series F: Computer and Systems Sciences, Vol. 151

Springer

Berlin
Heidelberg
New York
Barcelona
Budapest
Hong Kong
London
Milan
Paris
Santa Clara
Singapore
Tokyo

Computational and Conversational Discourse

Burning Issues – An Interdisciplinary Account

Edited by

Eduard H. Hovy

Information Sciences Institute, University of Southern California
4676 Admiralty Way, Marina del Rey, California 90292-6695, USA

Donia R. Scott

Information Technology Research Institute, University of Brighton
Lewes Road, Brighton BN2 4NT, UK

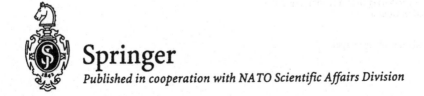

Springer
Published in cooperation with NATO Scientific Affairs Division

Proceedings of the NATO Advanced Research Workshop on Burning
Issues in Discourse, held in Maratea, Italy, April 13–15, 1993

Library of Congress Cataloging-in-Publication Data

Computational and conversational discourse : burning issues, an
  interdisciplinary account / edited by Eduard H. Hovy, Donia R.
  Scott.
      p.   cm. -- (NATO ASI series. Series F, Computer and systems
  sciences ; no. 151)
   Includes bibliographical references and index.

   1. Discourse analysis. 2. Computational linguistics.   I. Hovy,
  Eduard H. II. Scott, Donia R., 1952-   . III. Series.
  P302.C6215 1996
  401'.41--dc20   96-27091
                                                          CIP

CR Subject Classification (1991): J.4, I.2, I. 7

ISBN 978-3-642-08244-3

© Springer-Verlag Berlin Heidelberg 2010
Printed in Germany

Printed on acid-free paper

Preface

We live in a world of interaction. Something happens when two people sit down next to one another in a bus and start talking; when someone picks up and reads a book that was written centuries earlier; when a politician or preacher broadcasts an appeal that sways a nation; when a mother and a child read a story together. The peculiar interplay of real-world facts, personal memories, imagination, emotions, interpersonal role-play, medium of communication, and background situation, is so complex that we cannot even categorize the major aspects involved, yet so natural that every toddler can do it, and *needs* to do it – whatever 'doing it' is. The ceaseless drive to communicate with one another seems almost a biological imperative: it is often harder *not* to communicate, even with a group of total strangers, than it is to perform (at the minimum) some greeting and rudimentary interaction.

But given the pervasiveness of extended turn-taking communication, our knowledge of how discourse works is amazingly sketchy. We have only rudimentary models of important phenomena such as conversational initiative and turn-taking; we understand very little about the processes of understanding and interpreting sentences, or about the mental structures that support these processes; we have no adequate means of defining even such basic building blocks of communication as the words "however" or "consequently".

The reasons for our ignorance are clear. Language is rooted in almost everything we experience, think and do, as individuals and as social animals; an adequate account of discourse will have to employ concepts from Anthropology and Sociology (to explain the interlocutors' group interactional behavior), from Linguistics (to explain grammar and lexis), from Neurolinguistics (to explain errors and quirks), from Semantics, Logic, and Philosophy (to describe the knowledge involved in communications), and from Artificial Intelligence and Cognitive Psychology (to describe the processes by which communication takes place), at the very least. Consequently, a full account of how language works is impossible – it would require a full account of human life!

This does not mean that we should give up, of course. It simply means that we should be willing to diversify our investigations; that people should study language from the points of view of Linguistics, of Anthropology, of Logic, and so on, independently. This almost certainly means that the insights reached from such different approaches, and even the methodologies of study employed by the different approaches, will tend to diversify. And while this is a good thing – only by such a multi-pronged study can we hope not to miss important aspects – unchecked diversification has its dangers: fragmentation and isolation.

It is consequently important and useful for researchers of the same, or closely related, phenomena of language to meet and compare their ideas, insights, methodologies, and results. Hopefully, some cross-fertilization may take place. Possibly, even, some new collaborations may be established.

Such cross-informing and fertilization was the goal of a workshop out of which this volume grew. *Burning Issues in Discourse* was held in Maratea, Italy, in April 1993, with funding generously provided by NATO and the Association for Computational Linguistics. The approximately thirty-five invited speakers and attendees at the workshop included researchers from around the world working in the areas of Computational Linguistics, Psycholinguistics, Text Linguistics and Sociolinguistics. The intention of the workshop was not to produce a grand new theory but rather to inform one another about the facets of the problems of discourse and about available methods of addressing them.

Since no comparable workshop about discourse had ever been held, the speakers were asked to select a specific topic and to approach it in a way characteristic of their field. In this way, participants would have a chance not only to learn about aspects of discourse possibly new to them, but also to experience, first-hand, how an eminent researcher from another field goes about his or her business.

The success of this workshop, and the richness and diversity of ideas presented, prompted this book. The editors solicited papers from all attendees, and after a stringent review of more than 20 papers submitted, a representative set was selected and collected as this volume. Some time elapsed between the workshop and the publication of the book. However, given that research methodologies do not change quickly, and that the various papers discussed issues with ongoing impact, the book embodies a contribution of longstanding interest.

This volume is intended to capture aspects of the study of discourse toward the end of the 20th century, in a variety of disciplines. Of interest are not only the analyses and results listed in the papers, but perhaps more importantly, a cross-disciplinary comparison of the questions asked and the methodologies adopted.

Seven papers are included: from Sociology, a paper by Schegloff; from Linguistics, papers by Martin, Ono and Thompson, and Hajicová from Computational Linguistics, papers by Dahlgren and Hobbs; and taking an empirical, experimental approach, a paper by Passonneau and Litman.

Questions addressed in these papers cover several distinct ranges of topics:

- from the different types of meaning simultaneously communicated in a discourse, with appropriate representations (Martin), to communicative intentions and inference processes, with representations of them (Hobbs);
- from what is said, and not said, in multi-party interaction (Schegloff), to how what is said is embodied in syntax (Ono and Thompson);

- from the internal structure of discourse (Dahlgren, Passonneau and Litman), to external signals of discourse structure (Hajicová).

The methodologies adopted also span a wide range, from hypothesis and analyses remaining to be verified experimentally (Martin, Schegloff), to tightly controlled and measured studies of human behavior in very specific settings and tasks (Passonneau and Litman), with computer studies (Hobbs, Dahlgren) and comparison or substitution studies (Ono and Thompson, Hajicová) in between.

It will be noted, in general, that as the 'scope' of the phenomena studied increases (from words to paragraphs to texts), the precision of the descriptive notation employed decreases. At the broad end of the scale, an ethnomethodological description of the patterns of reasoning underlying certain interactions (Schegloff) is performed using the natural language English; at the opposite, highly focused end, a discussion of the step-by-step inference processes required to interpret single utterances (Hobbs) is performed using first and higher-order predicate calculus. These two papers, with Martin's explicit exploration of how various notations are suited to capture various aspects of discourse, illustrate a fundamental tension in the study of discourse: while being precise and very detailed, a formalism such as Hobbs's calculus quickly becomes opaque and difficult to interpret when more than a few clauses at a time are studied, thus limiting the scope of practical application of the ideas. On the other hand, the flexibility of well-written prose such as Schegloff's and Martin's allows discussion of extremely complex and subtle phenomena, without however providing specific additional hints about the generalizations, processes, etc., that underlie the phenomena in general. This is hardly surprising. As in any other intellectual enterprise, our task is to develop more concise, precise, and intuitive notations (including primitives such as 'initiative', 'discourse segment', 'communicative goal', as well as formalisms for discourse structures and rhetorical relation schemas). These tools provide levers with which we can pry loose the layers of the mystery of discourse. Here Martin's metaphors of the particulate, periodic, and prosodic nature of aspects of discourse is an interesting attempt to capture phenomena beyond the rather limited range usually studied by people trained in the mathematical/logical tradition. One step closer to formalization are the discourse structure relations postulated by Mann and Thompson, Hobbs, Dahlgren, and others, as described in Dahlgren's paper, as well as the idea of discourse segments, as postulated by Grosz and Sidner and others, and described in Dahlgren and tested by Passonneau and Litman.

It is important to bear in mind that cross-disciplinary discussion is one of the most effective mechanisms by which insights progress from metaphors, via provisional term systems, to formal, well-defined notations. Schegloff's 'action-by-inaction' and Ono and Thompson's embodiment of meaning in syntactic forms might eventually be incorporated into a formal system, in a way similar to the incorporation of certain discourse structure (rhetorical) relations into the formal system of Kamp's DRT, described in Dahlgren's paper. This is not to imply that it is the necessary path of ideas to proceed from English descriptions toward first-order (or any other kind) of predicate logic, or to imply that ideas are mature only when they can be expressed in such logics; logic is simply one method to write down symbol systems, neither the only nor necessarily the most suitable one for the phenomena of discourse. But this view does adhere to the belief that a model of a complex phenomenon is most complete, and comes closest to a theory in the strict sense of the word, when it provides a notation and a taxonomization of phenomena in which precise predictions can be made, tested, and verified or falsified.

In the large body of literature available on discourse as an object of study, a certain set of issues has been discussed a great deal. Some of these issues are directly or indirectly addressed by papers in this book. These include:

- discourse is structured; in particular, blocks of utterances are grouped into segments. However, discourse structure is not amenable to the methods of single-sentence grammatical analysis: there are no 'discourse grammars';
- multi-party dialogues involve initiative signaling and negotiation;
- in interactions, several kinds of meaning are communicated simultaneously, including the argument structure ('the point'), factual or imagined details ('the content'), and the effects of the communication medium and setting.

In more detail, these issues can be separated out as follows.

Multi-Party Discourse: The collaborative construction of a coherent discourse involves several factors that complicate the single-speaker picture, for example, turn-taking, signalling and negotiating initiative. How well do current theories account for these phenomena? Can they be used in computational systems? What needs to be added, and how can the open questions be addressed in testable ways? Two papers address questions in this area: the ways people communicate by not saying anything out loud (Schegloff), and the ways people structure their message when they do say something (Ono and Thompson).

Discourse Segmentation: Coherent discourse is structured. What does this structure look like (e.g., trees vs. networks; single vs. multiple structures)? How are the structural segments defined? What are the relevant units of segmentation: propositions, sentences, utterances? How are their boundaries signalled? What is the nature of intersegment relations—is it intentional, semantic, structural, or all three? What role does intentionality play in the segmentation? The computational linguist Hobbs addresses communicative intention and its effects on discourse structure, while Dahlgren reviews various theories of discourse structure and intersegment relations. Passonneau and Litman empirically study people's ability to segment discourses the same ways.

Information in Discourse: What type(s) of information can be communicated by discourse? How? Martin provides an analysis of several texts, arguing that at least three different mechanisms are used in discourse to communicate three fundamentally different kinds of meaning, in parallel. Looking within each discourse segment, and with each clause, more questions arise. Information is not presented randomly within discourse segments, and segments themselves are not randomly ordered. What governs the flow of information? What is the difference between notions such as Topic, Theme, Focus, and Given? How does information presentation (by the speaker) influence information access (of the hearer)? From the perspective of the Prague School, Hajicová describes some of the pertinent distinctions between Topic and Focus.

Discourse Structure and Syntactic Form: How do discourse and syntactic structures relate? How do they constrain one another? How can one identify correlations between them and specify the correlations as rules for, say, automated discourse generation? In their paper, Ono and Thompson describe the syntactic means used to communicate specific intentions and ideas, while Hajicová describes various syntactic means for expressing focus. Martin extracts various types of meaning from syntactic and lexical cues.

Tools, Techniques, and Experimental Methodologies: How can theories of discourse be empirically verified? All the abovementioned topics can benefit from the development and application of objective testing techniques. What techniques and methodologies exist? What aspects of discourse do they best address? In his Postscript, Schegloff pointedly discusses aspects of the methodology employed in his investigations. The study by Passonneau and Litman measures agreement in people's segmentation of discourses, and describes the cues used by people. Dahlgren mentions corpus studies that identified and analyzed segment cue words.

We hope the reader gets as much from reading the book as we did from editing it.

June 1996

Eduard H. Hovy
Information Sciences Institute
University of Southern California

Donia R. Scott
Information Technology Research Institute
University of Brighton

Table of Contents

XII   Table of Contents

Perspective from Sociology

Chapter 1
Issues of Relevance for Discourse Analysis: Contingency in Action, Interaction and Co-Participant Context

Emanuel A. Schegloff
University of California in Los Angeles

1. Introduction

There are three themes on which I would like to focus attention, whose full incorporation into the analysis of discourse is, in my view, critical for its optimum further development. What needs to be incorporated is an orientation 1) to action, 2) to interaction, and 3) to multi-party interaction. It will turn out that orientation to each of these themes confronts the student of discourse with a sort of challenge whose depth and consequentiality has not yet been fully registered or explored, but is likely to be substantial. What becomes inescapable in facing up to action, interaction and multi-party interaction is the challenge of contingency. What exactly I mean by "contingency" will only come into view over the course of the discussion of empirical materials; as it cannot be usefully elaborated here, I will return to the import of contingency at the end.

But before launching into this agenda, I need to make clear several premises of what I have to say—both as context for my central points and to make explicit my understanding of discourse's place in the world.

The first is that I take real world, naturally occurring ordinary discourse as the basic target; it is as a student of that that I offer what follows. There may well be grounds for those with other interests to opt for a different point of reference or a different target of inquiry; but for me these involve departures from the natural and cultural bedrock.

Second, whereas for many linguists and other students of language, conversation is one type or genre of discourse, for me discourse is, in the first instance, one kind of product of conversation, or of talk-in-interaction more generally.[1] It can be a contingent product of participants in ordinary conversation;

---

[1] The term "discourse" now has a variety of uses. In contemporary cultural criticism one can speak of the "discourse of modernity" or "the discourses of power" or "feminist discourse;" indeed, I was tempted to begin the present sentence by referring to "the discourse of contemporary cultural criticism." In a more technical usage current among linguists and computational linguists, as one reader has reminded me, "...'discourse' is simply a broad term that includes interactional talk, but also includes written essays, advertisements, sermons, folk tales, etc. With this view of 'discourse,' your characterization

or it can be the designed product of a form of talk-in-interaction which is some systematic variant or transformation of ordinary conversation—like the interview or the lecture. But I take conversation to be the foundational domain. And this leads to the last point of departure I want to make explicit.

I take it that, in many respects, the fundamental or primordial scene of social life is that of direct interaction between members of a social species, typically ones who are physically co-present. For humans, talking in interaction appears to be a distinctive form of this primary constituent of social life, and ordinary conversation is very likely the basic form of organization for talk-in-interaction. Conversational interaction may then be thought of as a form of social organization through which the work of the constitutive institutions of societies gets done—institutions such as the economy, the polity, the family, socialization, etc. It is, so to speak, sociological bedrock. And it surely appears to be the basic and primordial environment for the development, the use, and the learning of natural language.

Therefore, it should hardly surprise us if some of the most fundamental features of natural language are shaped in accordance with this home environment in co-present interaction—as adaptations to it, or as part of its very warp and weft (Schegloff, 1989, 1996). For example, if the basic natural environment for sentences is in turns at talk in conversation, we should take seriously the possibility that aspects of their grammatical structure, for example, are to be understood as adaptations to that environment. In view of the thoroughly local and interactional character of the deployment of turns at talk in conversation (Sacks, Schegloff and Jefferson, 1974), grammatical structures—including within their scope discourse—should in the first instance be expected to be at least partially shaped by interactional considerations (Schegloff, 1979).

---

is hard to interpret." My point is meant to contrast with this fundamentally taxonomic usage. The taxonomic usage reflects academic interests in discriminating and conceptualizing a variety of genres, and the relationship of these genres is derived from their relative positioning in this conceptuial mapping, not in the naturally occurring processes which might conceivably have engendered them. It is this which the point in the text is meant to invoke. That point turns on what is both a broader and a narrower sense of "discourse," one which underlies these other usages (and is a common characteristic of the usages discussed in the Oxford English Dictionary), and that is the usage which contrasts "discourse" with single sentences. If one examines the usage of a term like "discourse analysis," for example, one rarely finds it invoked to deal with single sentences. "Discourse" regularly refers to extended, multi-sentence "texts." And it originally had reference to speech or talk. Hence my point, which is that discourse—extended or multi-unit talk production—be understood processally—as one sort of product of conversation, rather than conversation being understood taxonomically, as simply one sub-type of discourse. In this view, extended spates of "text" by a single speaker have as their source environment turns-at-talk in conversation in which that is the concerted product of a company of participants in interaction, for example, spates of story-telling. A kind of virtual natural history of interactional genres and speech exchange systems may then track the disengagement of such sustained, mutli-unit talk production by a single speaker from the interactional environment of conversation into settings such as religious ceremony, political speech making, prophetic invocation, philosophical disquisition, etc., and the development of writing then enables an explosion of yet further genres.

So much for premises. The three themes on which I wish to focus your attention are endemic to the organization of talk-in-interaction, and follow from these points of departure. The first concerns the centrality of action.

2. Action

Among the most robust traditional anchors for linguistic analysis beyond the level of syntax are orientations to information and truth. This position needs to be reconsidered. It is critical that the analysis of discourse incorporate attention not only to the propositional content and information distribution of discourse units, but also to the *actions* they are doing.[2] Especially (but not exclusively) in conversation, talk is constructed and is attended by its recipients for the action or actions which it may be doing. Even if we consider only declarative-type utterances, because there is no limit to the utterables which can be informative and/or true, the informativeness or truth of an utterance is, by itself, no warrant or grounds for having uttered it—or for having uttered it at a particular juncture in an occasion. There is virtually always an issue (for the participants, and accordingly for professional analysts) of what is getting *done* by its production in some particular here-and-now.

Although I cannot undertake here to go beyond asserting this to demonstrate it, I do want to exemplify it. In order to make vivid the consequentiality for conversational participants of the action which an utterance is doing, quite apart from the information which it is conveying, I offer a condensed and partial analysis of one conversational fragment. I hope thereby to show at least one way that action can matter, and indicate an order of analysis which this field of inquiry must incorporate if this view of the inescapability of action is correct.

In the conversation between Debbie and Nick (who is her boyfriend Mark's roommate) which is reproduced in its entirety in Appendix A (at whose end the most pertinent notational transcription conventions are explained), a peculiarly insistent exchange develops which can serve to exemplify my theme.

```
(1)   Debbie and Nick:34-69

34    Debbie:    ·hhh Um:: u- guess what I've-(u-)wuz lookin'
                 in the paper:.-
35               -have you got your waterbed yet?
36    Nick:      Uh huh, it's really nice °too, I set it up
37    Debbie:    Oh rea:lly? ^Already?
38    Nick:      Mm hmm
39               (0.5)
40    Debbie:    Are you kidding?
```

---

2 As will become clear below, I do not mean here to be invoking speech act theory, whose ability to deal with real ordinary discourse is subject to question, but that is another story (cf. Schegloff, 1988a, 1992a, 1992b:xxiv–xxvii).

```
41    Nick:      No, well I ordered it last (week)/(spring)
42               (0.5)
43    Debbie:    Oh- no but you h- you've got it already?
44    Nick:      Yeah h! hh=                     ((laughing))
45    Debbie:    =hhh [hh ·hh]                    ((laughing))
46    Nick:           [I just] said that
47    Debbie:    O::hh: hu[h, I couldn't be[lieve you c-
48    Nick:               [Oh (°it's just) [It'll sink in
49               'n two day[s fr'm now (then    ) ((laugh))]
50    Debbie:              [      ( l a u g h ) ) ]
              Oh no cuz I just got- I saw an ad in the
51            paper for a real discount waterbed s'
52            I w'z gonna tell you 'bout it=
53    Nick:      =No this is really, you (haven't seen)
54               mine, you'll really like it.
55    Debbie:    Ya:h. It's on a frame and everythi[ng?
56    Nick:                                        [Yeah
57    Debbie:    ·hh Uh (is) a raised frame?
58    Nick:      °mm hmm
59    Debbie:    How: ni::ce, Whadja do with Mark's cou:ch,
60               (0.5)
61    Nick:      P(h)ut it out in the cottage,
62               (0.2)
63    Nick:      goddam thing weighed about two th(h)ousand
              pound[s
64    Debbie:        [mn:Yea::h
65               I'll be[:t
66    Nick:             [ah
67               (0.2)
68    Debbie:    Rea:lly
69               (0.3)
```

At a point which I will characterize in a moment (35), Debbie asks Nick whether he has gotten his waterbed yet. He tells her that he has, and this is met with three rounds of questioning, challenging, or disbelief—to settle for pre-analytic characterizations initially. First, (at 37) "Oh really? Already?" When Nick confirms, she asks again (40), "Are you kidding?" "No," he says, and notes that it has been a while since he ordered the waterbed. And still again she asks (43) "Oh no but you h- you've got it already?" Finally, Nick complains (46) that he has already said so. What is going on here?

Debbie has asked a seemingly simple, informational question, and Nick has answered it. Now questioning of the sort which Debbie engages in here can be undertaken in conversation (among other uses) as a kind of harbinger of disagreement—sometimes verging on challenge, and one response to such a usage is a backdown by its recipient. Sometimes this is a backdown in the substance of what was said,[3] sometimes in the epistemic strength with which it was put

---

3 For example,

```
      A:    Is Al here?
      B:    Yeah
            (0.?)
```

forward.[4] If a first questioning does not get such a backdown, sometimes a second one does. But what kind of backdown is possibly in order here? If Nick has in fact taken possession of his waterbed, is he now to deny it? Is he to retreat to a position of uncertainty or supposition about the matter? What could Debbie be after?

It is also true that, in keeping with the peculiar interactional "style" of teasing and laughing which some Americans in their late teens and early 20's practice, Nick has been indulging himself in unrelieved "kidding around" in the earlier part of this conversation, and it is not implausible that, if the first of Debbie's responses was hearably "surprise," the second could be checking out whether this is not just more teasing by Nick. But then what is the third about (at 43)? And why the persistence of her stance? Why should this information come in for such scrutiny and doubting?

We can get analytic leverage on what is going on here if we attend to these utterances not only as a matter of information transfer involving issues of truth and confidence, but as actions in a course of action, constituting an interactional sequence of a recurrent form. For it is not enough that a speaker has something to tell and undertakes to tell it; the prospective teller must find a way to tell it, and that implicates a recipient as well, a proper recipient aligned to recipiency, and not simply a sentient body with functioning eyes and ears. And this involves action in interaction.

```
     C:   He is?
     B:   Well he was.
```
4 For example, in the following fragment from a conversation in a used furniture store (US, 27:28-28:01), Mike is angling to buy (or be given) Vic's aquarium when Rich intervenes with a challenge to Vic's ownership of it (at line a). Note the backdowns in epistemic strength at lines c and e in response to Vic's questionings at lines b and d respectively—first from assertion to assertion plus tag question, and then to fully interrogative construction. (Note finally that in the end Vic does disagree with Rich's claim.)

```
     MIK:  Wanna get some- wannuh buy some fish?
     RI?   Ihhh ts-t
     VIC:  Fi:sh,
     MIK:  You have a tank I like tuh tuh- I-I [like-
     VIC:                                     [Yeh I gotta fa:wty::
           I hadda fawtuy? a fifty, enna twu[nny:: en two ten::s,

     MIK:                                    [Wut- Wuddiyuh doing wit
           [dem. Wuh-
a    RIC:  [But those were uh::::[Alex's tanks.
     VIC:                        [enna fi:ve.
b    VIC:  Hah?
c    RIC:  Thoser' Alex's tanks weren't they?
d    VIC:  Pondn' me?
e    RIC:  Weren't- didn' they belong tuh Al[ex?
     VIC:                                    [No: Alex ha(s) no tanks
           Alex is tryintuh buy my tank.
```

Begin by noting (at 34) Debbie's "guess what." This is a usage virtually dedicated to a particular type of action referred to in past work as a "pre-announcement" (Terasaki, 1975). Announcements, or other prospective "tellings," face the familiar constraint that they generally should not be done to recipients who already know "the news." Pre-announcements and their responses—pre-announcement sequences, that is—allow a prospective teller and recipient to sort out together whether the "news" is already known, so that the telling or announcement can be withheld or squelched, if need be. Of course, the very doing of a pre-announcement displays its speaker's supposition that there is indeed news to tell, and to tell as news to this recipient. Still, one thing prospective tellers can do (and regularly do do) before telling is to check whether the news is already known. And among the recurrent response forms to such pre-announcements, two central types are the "go ahead" type of response (such as, in response to "guess what," "what"), which forwards the sequence to its key action—here announcing or telling, and the "blocking" type of response (for example, a claim of knowledge, such as "I heard"), which aims to forestall such telling.[5]

Often the pre-announcement provides clues about the news to be told (e.g., "Y'wanna know who I got stoned with a few weeks ago?," or "You'll never guess what your dad is looking at," Terasaki, 1976:27–28), the better to allow the recipient to recognize it if it is already known, and to provide a context for understanding it and an interpretive key, if it is not already known. And here Debbie does provide such clues; "I was looking in the paper" (at 34) intimates that what she has to tell is something that one can find (and that she has found) in the newspaper. And then (at 35), "have you got your waterbed yet?" So the thing to be told (about) has something to do with waterbeds, and Nick's possibly being in the market for a waterbed in particular.

So there is another constraint on Debbie's telling here, one which is not generic to "telling" in the way in which "already known-ness" is. Debbie has information to offer—information which is relevant to Nick only contingently. Offers and offer sequences too can take what we call "pre-sequences," just as announcements can and do. With pre-offers, prospective offerers can try to assess whether what they have to offer is relevant to their recipients and may be welcomed by them, so as to not make offers which will be rejected, for example. What Debbie has to offer is information on a cheap waterbed or an especially desirable one, but her pre-offer is designed to find out whether such information is relevant to Nick— whether what will be offered will be relevant. That is what "Have you got your waterbed yet?" appears designed to do—it is an analyzable pre-offer.[6] As such, it

---

[5] For a more general treatment, cf. Terasaki, op. cit.; Schegloff, 1990. For an instance with both—indeed, simultaneous—go-ahead and blocking responses, Schegloff, forthcoming.

[6] Among the design features which make it so analyzable is the negative polarity item "yet," which displays its speaker's orientation to a "no" answer, and builds in a preference for that sort of response (note that "yet" is replaced by "already" after Nick's affirmative response). The placement of the pre-offer *after* the pre-announcement is a way of showing the former to be in the further service of the latter, and part of the same "project." For a formally similar series of sequences, see the data excerpt in footnote 14, where positioning

too (like pre-announcements) takes among its alternative response types a go-ahead response—which forwards the sequence to an offer, and a blocking response which declines to do so.

So when Debbie asks, "Have you got your waterbed yet?" she is not just asking for information; she awaits a go-ahead to the pre-offer, on which her offer of the information which she has come across in the newspaper has been made contingent. And when Nick responds affirmatively, he is not only confirming the proposition at issue—that he already has his waterbed; he is blocking her from going on to tell the information which she has seen in the newspaper.

And this is the proximate sequential and interactional context for Debbie's repeated questionings. The backdown which is relevant here concerns not the facticity of the presence of a waterbed, and not Nick's confidence in asserting it; and perhaps not even whether he is teasing. What is at issue is a backdown from the blocking response to the pre-sequences. One form it could take is, "why?" As in (starting at 37–38) "Oh really? Already?" "Mm hmm, why?" Or (at 40), "Are you kidding?" "No, why? Or (at 43–44), "Oh- no but you h- you've got it already?" "Yeah! Why?"

As it happens, it appears that Nick has not caught this, and so he responds only at the level of information transmission.[7] When for the third time Debbie asks, "You've got it already?" he says, "Yeah, I just said that...It'll sink in in one or two days from now." That is, he just says it again—and more pointedly; he makes her out to be not too quick on the uptake; she'll get it eventually.[8]

But it is he who has apparently not gotten it. And it will be we who do not get it if we do not systematically distinguish what an utterance is about or what is it saying, on the one hand, from what it is doing on the other. Backing down from one is quite different from backing down from the other. Attention will virtually always need to be paid to the issue "what is someone doing with some utterance? What action or actions are involved?" Because overwhelmingly actions are involved, they are oriented to by the participants both in constructing and in understanding the talk, and the discourse cannot be appropriately understood without reference to them—precisely because they are key to the participants' conduct.

It follows, of course, that the actions to which analysis needs to attend are not classes of action defined by the conceptual commitments of professional discourse analysts (as, for example, in any of the varieties of academic speech act theory), but those units and understandings of action which are indigenous to the actors'—the interactional participants'—worlds. Hence, the appearance in my account of actions like "pre-offer" or "pre-announcement," which figure in no speech-act

---

"Didjer mom tell you I called the other day?" after "Wouldju do me a favor?" puts it under the jurisdiction of the projected request sequence, and in pursuit of that project.

[7] It is possible, of course, that he *has* caught it, but prefers not to hear of the better buy he could have had, having just taken possession of, and pride in, his new acquisition.

[8] Let me just mention without elaboration that Debbie does find a way of conveying what she saw in the newspaper in spite of it all, namely, in the questions she eventually asks about Nick's waterbed—specific questions (about the bed being on a frame, on a raised frame, etc., cf. Lines 55–57) almost certainly prompted by what she saw in the paper.

theory with which I am familiar, but exemplars of which are laced through and through ordinary conversation.

That is the first theme I want to put before you: how an action done by a speaker—taken as an action—has decisive consequences in shaping the trajectory of the talk's development.   The second theme concerns how the absence of an action can have such consequences. But the absent action here is not that of the speaker of the discourse but rather of its recipient, and this forces on us the issue of the interactivity of discourse production.

3. *Inter*action

When I say that the second theme on which I want to focus is *inter*action, I should note that what I mean by the term "interaction" here may be the same as what some workers in the area of discourse analysis and computational linguistics mean by this term. For students of interaction, "multi-party" means "more than *two*." For at least some students of discourse, it apparently means "more than *one*." And "produced by more than one" is what I mean here by the *interactive* production of discourse.

It is some fifteen years now since Charles Goodwin (1979, 1981) gave a convincing demonstration of how the final form of a sentence in ordinary conversation had to be understood as an interactional product. He showed that the speaker, finding one after another prospective hearer not properly aligned as an actual recipient (that is, not looking at him),   re-constructed the utterance in progress—the sentence—so as to design it for the new candidate hearer to whom he had shifted his gaze. He showed the effects on the utterance of both the candidate recipients' conduct and the speaker's orientation to the several possible recipients—a feature we call recipient design. Goodwin's account served at the time (and still serves) as a compelling call for the inclusion of the hearer in what were purported to be speaker's processes, and for the inclusion of the non-vocal in purportedly vocal conduct. In a paper published the following year, Marjorie Goodwin (1980) provided another such demonstration, showing how a hearer's displayed uptake and assessment of a speaker's in-process talk shaped the final form which the utterance took.[9]

The general point here is that units such as the clause, sentence, turn, utterance, discourse—all are in principle *interactional* units. For it is not only that turns figure in the construction of sequences (by which I mean action sequences implemented through talk and other conduct). Sequences—and their projected, contingent alternative trajectories—figure in the construction of turns, and of the

---

[9] Others have contributed to this theme as well. I leave with a mere mention Lerner's work (1987, 1991, frth), pursuing several observations by Sacks (1992:I:144–147 et passim; 1992:II:57–60 et passim), on "collaboratives," in which two or more speakers collaborate in producing a turn, in the sense that each actually articulates part of it. See also Schegloff, 1982, 1987, Mandelbaum, 1987, 1989, and in a somewhat different style of work, the papers in Duranti and Brenneis, 1986 and Erickson (1992).

extended turns which we sometimes call discourse(s). In examining the following conversation, I want to explicate how the sequence which is being incipiently constructed figures in the production of what appears to be an extended spate of talk by a single speaker—a discourse of sorts.[10]

(2)   Marcia and Donny: Stalled

```
01    1+ rings
02    Marcia:    Hello?
03    Donny:     'lo Marcia,=
04    Marcia:    Yea[:h        ]
05    Donny:     =[('t's) D]onny.
06    Marcia:    Hi Donny.
07    Donny:     Guess what.hh
08    Marcia:    What.
09    Donny:     ·hh My ca:r is sta::lled.
10               (0.2)
11    Donny:     ('n) I'm up here in the Glen?
12    Marcia:    Oh::·.
13               {(0.4)}
14    Donny:     {·hhh }
15    Donny:     A:nd.hh
16               (0.2)
17    Donny:     I don'know if it's po:ssible but {·hhh/0.2}
18               see I haveta open up the ba:nk.hh
19               (0.3)
20    Donny:     a:t uh: (·) in Brentwood?hh=
21    Marcia:    =Yeah:- en I know you want- (·) en I whoa-
22               (·) en I would, but- except I've gotta leave
23               in aybout five min(h)utes.[(hheh)
24    Donny:                               [Okay then I
25               gotta call somebody else.right away.
26               (·)
27    Donny:     Okay?=
28    Marcia:    =Okay [Don    ]
29    Donny:           [Thanks] a lot.=Bye-.
```

---

[10] The following discussion documents another point as well. A number of papers (e.g., Jefferson and Schenkein, 1978; Schegloff, 1980, 1988c,1990) describe various ways in which sequences get expanded as the vehicle for interactionally working out some course of action between parties to talk-in-interaction. Sequence expansion is embodied in the number of turns composing the trajectory of the sequence from start to closure. But the amount of talk in a sequence can increase in ways other than expansion in its sequence structure. Among these is expansion of the component turns that make up the sequence. (Cf. Zimmerman, 1984:219–220 and the discussion in Schegloff, 1991:62–63 concerning different formats of citizen complaint calls to the police.) Most commonly it is the second part of an adjacency-pair based sequence which gets this sort of elaboration, as when a question gets a story or other elaborated response as its answer. There may then still be a "simple," unexpanded (or minimally expanded) sequence structure of question/answer, or question/answer/receipt, with the second of these parts being quite a lengthy "discourse unit." "Turn expansion" may then stand as a contrast or alternative to sequence expansion, rather than in a subsuming or subsumed relationship to it (compare Schegloff, 1982:71–72). In the data examined in the next portion of the text, the discourse or turn expansion occupies not the second part position in the sequence, but the first.

30   Marcia: Bye:.

The "discourse of sorts" which eventually gets produced here (at lines 9, 11, 15, 17–18, and 20) could be rendered as follows:

> My car is stalled (and I'm up here in the Glen?), and I don't know if it's possible, but, see, I have to open up the bank at uh, in Brentwood?

Put this way, each component (e.g., each clause or phrase) appears to follow the one before it, although I have tried to capture with punctuation the possibly parenthetical character of the second component, with attendant revised understanding of the relative organization of the components surrounding it. Now aside from the "Oh" interpolated by Marcia (at line 12) in response to this element, all that I appear to have left out in this rendering of the talk is . . . *nothing*—that is, silences, some of them filled by hearable in- and out-breaths. But, of course, these silences are *not* nothing. The something that they are—the something that each is—is given by its sequential context, and it is *that* which requires us to attend to the actions being done here . . . and *not* being done here. Then we can see that—and how—this is not a unitary discourse produced by a single participant, and how some of its components follow not the components which preceded them, but the silence which followed the component that preceded them. Thereby we can come to see that it is not just a hearer's uptake and actions which can enter into the shaping of a speaker's talk; it can be the absence of them which does so.

To begin then, the utterance at line 07 should now be readily recognizable for the action which it is doing: it is (doing) a pre-announcement. It may be useful to be explicit about what is involved in making and sustaining such a claim. Virtually always at least two aspects of a bit of conduct—such as a unit of talk—figure in how it does what it does: its position and its composition (Schegloff, 1992c:1304–1320). A sketch will have to suffice.

We have already noted that this formulaic utterance "Guess what" is virtually dedicated to doing pre-announcements, as are various extensions and variants of it, such as "Guess what I did today," "Guess who I saw," etc.[11] This account of composition is only rarely available; precious few configurations of talk are so dedicated, and even those that are are contingent on their position. "Hello," said upon tripping over a prone body in a British film, is not a greeting, however much that expression might appear dedicated to doing so.

And what is the position of this utterance? How is it to be characterized? It comes just after the opening—the telephone ring's summons and the recipient's response (01–02), and the exchange of greetings intertwined with the explication

---

[11] Cf. Terasaki, op. cit. Note that such utterances are neither designed, nor are they heard, as commands or invitations to guess, i.e., to venture a try at what their speaker means to tell, though hecklers may heckle by so guessing (though I must say that I have seen very few empirical instances of this). On the other hand, some recipients of pre-announcements who know—or think they know—what the pre-announcer has in mind to tell may not simply block the telling by asserting that they know; they may *show* that they know by pre-empting the telling themselves.

of the identities of the two participants (03–06). I can only mention here something that would inform the parties' conduct of the ensuing interaction, namely the rushed, charged, almost breathless quality of Donny's participation, embodied here in his preemptive self-identification at line 5, rather than waiting to be recognized (Schegloff, 1979). It is a way of doing "urgency," and it is part of the positioning of "Guess what." Another part is the possible absence here of the start of an exchange of "Howaryou"s, a highly recurrent next sequence type in conversations between familiars under many (though not all) circumstances (Schegloff, 1986). In moving directly to "first topic" and the "reason for the call," Donny pre-empts "Howaryou"s as well, and this further informs the position in which "Guess what" is done. This position and the utterance in it, then, contingently foreshadow not only a telling of some news; they adumbrate the character of that news as well—that is, as urgent (or in some other respect "charged").

The pre-announcement projects further talk by its speaker, contingent on the response of the recipient, and we have already said a bit about the fairly constrained set of response types by the recipient which it makes relevant: a go-ahead response (the "preferred" one in the terminology of conversation analysis[12]), a blocking response, a preemptive response or a heckle-version of one. In the data before us, the response (at line 08) is a go-ahead.

The position (at line 08) is the turn after a pre-announcement which has made a response to it relevant next. The composition is common for responses to pre-announcements of the form "guess + question word" (as well as "y'know + question clause"): returning the question word from the pre-announcement ("Guess what." "What." "Y'know where I went?" "Where." etc.).[13]

With this response, Marcia both shows that she understands Donny's prior turn to have been a pre-announcement (thereby further grounding our analysis of it along these lines in the just preceding text), and provides an appropriate response to it. And note that that is how Donny hears Marcia's response; for otherwise, her "What" could invite treatment as displaying some trouble in hearing or understanding. It is not, of course, doing that, and it is not heard that way. "What" displays an understanding of "Guess what" as a pre-announcement; and Donny's ensuing turn displays his understanding of it as a go-ahead response to a pre-announcement. Of course Donny's ensuing turn—the one at line 09—is in the first instance otherwise engaged, and that is what we turn to next.

The pre-announcement sequence having been completed with a go-ahead, what is Donny's next utterance?

Well, in the first instance, it seems clearly enough designed to deliver the projected news. Note well: that it is conveying information is one formulation; that it does so by an utterance designed to be a recognizable action—"announcing," or "telling"—is another. For, of course, information can be

---

12 Cf. for example Heritage, 1984b:265-92; Levinson, 1983:332–356; Pomerantz, 1984; Sacks, 1987[1973]; Schegloff, 1988d:442–457.
13 Again, cf. Terasaki, op. cit. for a range of exemplars; Schegloff, 1988a.

conveyed by utterances designed to do something else in the first instance and on the face of it. But this one is clearly enough designed to do "telling."[14]

But what are the design features that make that "clear?" I can only tick off a series of observations whose development would be pertinent to such an analysis. First, the utterance is in an assertion or declarative format. Second, it refers to a speaker-specific event (what Labov and Fanshel called an "A-event"[15]). Third, it is presented as a recent, indeed as a current, event (Donny says "My car is stalled"). Fourth, as a current A-event, it is not otherwise accessible to recipient (by definition, else it would be an "A-B event") . There is undoubtedly more, and none of this may strike you as itself news. Still, if we are to get clear on how the actions which people do with talk "are" transparently what they "are," we will have to make analytically explicit how they are constructed to be transparently that (or equivocally that, for that matter), and how they may therefore be recognizable as transparently that (or equivocally that)—both to their recipients and (derivatively) to us as analysts.

It is not enough that there was a pre-announcement sequence with a go-ahead response. What follows is not necessarily an announcement; it will have to be constructed by its speaker as a recognizable, analyzable announcement, though its position after a pre-announcement sequence will potentiate such recognition. Once again, then: position and composition matter. So if discourse analysis takes the actions being done in the discourse as key to understanding its organization, then this will be part of the job.

Anyway, just as pre-announcements make sequentially relevant a response from some restricted set of next actions, so do announcements or tellings. Among them are some form of information uptake (such as registering the new information as new, for example through the use of the "oh" which Heritage (1984a) termed a "change-of-state token," or alternatively registering it as having already been known after all), or some form of assessment of what has been told— as good, awful, interesting, discouraging, etc. And indeed, these forms of action both regularly occur in the immediate sequential context of announcements. Not here, however.

---

[14] See, for example, Schegloff, 1990:63, footnote 6, for a discussion of the same bit of information first being conveyed in an utterance designed to do something else, and immediately thereafter done as a "telling" at arrows a and b respectively in the following exchange:

```
B:    But- (1.0) Wouldju do me a favor? heheh
J:    e(hh) depends on the favor::, go ahead,
B:    Didjer mom tell you I called the other day?          ← a
J:    No she didn't.
      (0.5)
B:    Well I called. (.) [hhh ]                             ← b
J:                       [Uhuh]
```

[15] By this they refer to "...representations of some state of affairs...drawn from the biography of the speaker: these are A-events, that is, known to A and not necessarily to B." Labov and Fanshel, 1977:62.

It now becomes pertinent for us to note that what follows this bit of news— "My car is stalled"—is silence, at line 10. Only two tenths of a second of silence to be sure; still, it is a silence after the prior speaker has produced a possibly complete utterance, one which makes relevant a response from its recipient, indeed, as noted, one which makes relevant quite specific types of response. Although everyone is silent (which silence as a state requires), someone in particular is "relevantly not talking," and that is Marcia. For Donny has produced a possibly complete turn, one which implicates some responsive action next—by Marcia. Absence of talk is then, in the first instance, attributable to Marcia. So although the effect of her silence is that no action seems to get done, what she is specifically and relevantly "not doing" is registering some uptake of what has been told, and/or some assessment of it—for it is these which Donny's announcement has made conditionally relevant.

At least that is some of what she is not doing. For a bit of talk can do more than one action. And some sorts of actions regularly serve as the vehicle or instrument by which other actions are done—announcements or tellings prominent among them (as are "questions" and "assessments"). In this case, I suggest, "My car is stalled" is not only an announcement, it is as well a complaint.[16]

The features which provided for this utterance as a possible "announcement" do not, of course, analyze its status as a possible complaint. In a variety of contexts it appears that formulating a state of affairs or an event as an absence, as a failure, as a non-occurrence is a way of constructing a recognizable complaint. And although the utterance under examination here is not as distinct an embodiment of such a usage in its surface realization as many others (for example, "You didn't get an ice cream sandwich," analyzed in Schegloff, 1988b:118–131), "stalled" is used to mean "engine will not start or run," i.e., it does formulate a failure.

Again, a complaint or report of trouble makes different types of response relevant next than does an announcement. Among such sequentially implicated next turns to complaints can be (depending on the character and target of the complaint or reported trouble) such ones as a sympathy expression, apology, excuse or account, agreement and co-complaint or disagreement and rejection, and—perhaps most relevant here—a remedy or help, or the offer of remedy or help.[17] So the silence at line 10 is to be understood not only for its withholding of news uptake and assessment, but for its withholding—by Marcia—of an offer to help. Though the silence by definition has no talk, it is as fully fledged an event in

---

[16] Alternatively, it could be characterized as a possible troubles telling (cf. Jefferson, 1988; Jefferson and Lee, 1981) or a pre-request (see below), though I cannot here take up the differences between these formulations, which in any case are not material to the issues I am presently concerned with.

[17] Drew (1984:137–139 et passim) describes the use of reportings which leave it to the recipient to extract the upshot and the consequent appropriate response. He addresses himself specifically to the declining of invitations by reporting incapacitating circumstances. His materials share with the present data the feature that a "dispreferred" action is circumlocuted by the use of a simple reporting of "the facts"—there declining invitations, here requesting a service.

the conversation as any utterance, and as consequential for the ensuing talk. The talk which follows is properly understood as following not the utterance "My car is stalled," not the information which that utterance conveys, and not the announcement which that utterance embodies or the complaint which that announcement implements; rather it follows the silence following that announcement, in which its "preferred" response (in the technical conversation-analytic sense of that term[18]) is hearably and analyzably withheld.

Note well: not every silence in conversation can be accorded an analysis along these lines. Silences get their interactional import from their sequential context (their "position"). A silence developing where an utterance has not been brought to possible completion is generally heard not as the interlocutor's, but as a pause in the continuing turn of the one who was talking (Sacks et al., 1974:715). And not all silences following a turn's possible completion are equivalent either: the silence following a question has a different import and consequence than one following an answer, or one following receipt of an answer. That something is missing, and what something is missing, should not simply be asserted; both need to be analytically grounded, based on structural analyses of relevant empirical materials. (This is so not only when silence develops, but at any apparent juncture in the talk where the analyst is drawn to introduce claims about what is "missing.")

Were sufficient space available, it would repay the effort to continue tracking in detail the development of this interaction, the whole of which lasts barely 18 seconds. A selective set of observation will have to suffice, focussing on the recurrent re-entries of Donny in the aftermath of "My car is stalled."

```
(3)   Marcia and Donny: Stalled (partial)

09    Donny:      ·hh My ca:r is sta::lled.
10                (0.2)
11    Donny:      ('n) I'm up here in the Glen?
12    Marcia:     Oh::.
13                {(0.4)}
14    Donny:      {·hhh}
15    Donny:      A:nd.hh
16                (0.2)
17    Donny:      I don'know if it's po:ssible but {·hhh/0.2}
18                see I haveta open up the ba:nk.hh
19                (0.3)
20    Donny:      a:t uh: (·) in Brentwood?hh=
21    Marcia:     =Yeah:- en I know you want- (·) en I whoa-
22                (·) en I would, but- except I've gotta leave
23                in aybout five min(h)utes.[(hheh)
```

Note to begin with that each of these re-entries (at lines 11, 15, 17 and 20) is constructed by Donny as an increment to the earlier talk, with the series of "turns-so-far" laced with silences, at many of which intervention from Marcia with an offer of help might be relevant. This incrementally constructed discourse is a

---

18 Cf. footnote 12.

whether an invitation will be accepted or declined, for example, is in principle indeterminate, much can be said about how either will be done if it is chosen—for example, whether it will be done promptly or delayed, explicitly or indirectly, baldly or with an account, etc. To be sure, that is also contingent, but there are orderly types of inferences which are observably generated if that type of next action is not done in that way—if, for example, an invitation is rejected precipitously, directly, explicitly and with no account. The co-participation of interlocutors in the production of talk, though a principled feature of talk-in-interaction, is always contingent in its occasioned expression. There are various places at which another can initiate talk and action, various practices for doing so, and (in multi-party interaction) alternative participants who can do so. But who, when and where are always contingent. There is virtually nothing in talk-in-interaction which can get done unilaterally, and virtually nothing which is thoroughly pre-scripted.

Contingency—interactional contingency—is not a blemish on the smooth surface of discourse, or of talk-in-interaction more generally. It is endemic to it. It is its glory. It is what allows talk-in-interaction the flexibility and the robustness to serve as the enabling mechanism for the institutions of social life. Talk-in-interaction is permeable; it is open to occupation by whatever linguistic, cultural, or social context it is activated in. It can serve as the vehicle for whatever concerns are brought to it by the parties engaging it at any given time. One underlying "burning" issue for computational interests in discourse analysis is how to come to terms with the full range of contingency which talk-in-interaction allows and channels. The themes of action, interaction and multi-party interaction on which I have focussed are three strategic—and I suspect under-appreciated—loci of this contingency.

Postscript

A referee of an earlier draft of this paper concluded a graciously appreciative assessment with a juxtaposition of its "whole method of analysis" to the referee's own "reservations," ones thought "likely to be shared by other readers," and suggested the possibility that "the author may want to explicitly address them in a preamble." I welcome the suggestion, though the reader will have noticed that I have preferred a post-amble, as it were, though I disavow the air of leisureliness which the neologism may hint at. If a reader shares the reservations, they will have been prompted by the paper, and should be addressed after that prompting, not before it. So here are the referee's reservations. I give them en bloc, and then take them up one by one.

> The problem for me is that the approach is purely descriptive and the analysis seems post hoc. The account of the data given by Schegloff seems persuasive, but is there any way of checking its validity? It seems that Schegloff is really setting up a hypothesis—(or really a set of hypotheses) that is interesting and plausible but which remains untested. My concern is

upward intonation, in the manner of a try-marked recognitional reference (Sacks and Schegloff, 1979) for a place, inviting its recipient's claim of recognition, and whatever other response might be forthcoming to this by now elaborately constructed, multiply laminated utterance.

Each of these increments comes after, and is analyzably directed to, the absence of any response to the complaint or (later) to the pre-request which Donny had presented as the reason for his call. When she eventually responds, Marcia declines to offer help, without ever saying "no." But her response does display (lines 21–22) her understanding that a solicitation of help was being made relevant ("en I know you want-") and that she would ordinarily comply ("en I would,"), but for a disabling circumstance.

Donny's "discourse of sorts," with the presentation of which this discussion began, has now been analyzed into the components from which it was assembled through a series of sequential and interactional contingencies, and its elaborate pursuit of help anatomized as the proposed underlying action. Here is one use of such analytic and terminological tools as the "parts" of an "adjacency pair," which are sometimes bemoaned as merely jargon. It is the analysis of "My car is stalled" as a possible announcement (a first pair part which makes one of a set of potential second pair parts relevant next), and consultation of other empirical announcement sequences (to establish what kinds of utterances serve as second pair parts which satisfy these sequence-organizational constraints), which grounds claims about what is missing in the following silence. It is analysis of that utterance as also a possible complaint (another type of first pair part), and examination of complaint sequences, that provides for the possible relevance next of the variety of responsive turn types proposed above, and characterizations of them as preferred or dispreferred, and underwrites further claims about what might be hearably missing. Without some such analytic resource (as well as analytic resources bearing on turn organization such as "possible completion" and further talk as either new "turn-constructional unit" or "increment" to the prior unit), it is easy for a post hoc observer (unlike an in situ participant) to overlook that an action is missing—precisely because the prior speaker (here Donny) may talk in such a manner as covers over and obscures that missingness, and makes it appear a mere pause in an ongoing utterance in progress. That action by the speaker, together with our vernacular inclination to normalize and naturalize the events in the interactional stream, can give the air of inevitability to what ends up having transpired. Stopping to say of "My car is stalled" that it is a possibly complete turn that is a first pair part, and what type or types of first pair part, prompts thinking explicitly about the possibly relevant second pair parts, prompts looking for them, and finding them "missing" if they are not there. The relevant "missing" is, if course, "missing for the participants," and one must then go back to the data to find evidence of an orientation to something being awry for the participants.

The point of this analysis, however, has been that not only is action a relevant facet and upshot of the talk, but that actions by other than the speaker are relevant to understanding a speaker's construction of discourse; and, relatedly, that the absence of actions by recipients—the absence of actions made relevant by the speaker's prior talk, the speaker's turn-so-far—may be crucial to understanding the speaker's further construction of the discourse.

This was my second "burning issue;" discourse involves not just action, but action in interaction, and the consequential eventfulness of its absence. Interaction, then, the relevant participation of a second party, the co-construction of discourse, may be most critical to our analysis of discourse when one of the participants is not producing talk—or doing anything else visible or hearable. For the very production of a discourse may be one contingent response by a prior speaker to the absence of a response by a co-participant to an apparently completed, action-implementing turn constructional unit.

4. *Multi-party* Interaction

My third theme concerns *multi-party* interaction. In light of the discussion so far, multi-party interaction can now be understood to refer to some instances of discourse in settings composed of more than two participants.[21] I will limit myself to sketching several organizational concerns which inform multi-party interactions and their participants which are not present (at least not in the same way) when there are only two participants. These concerns are relevant because they can enter into the design, implementation and understanding of the talk which composes the discourse, and a discourse analysis which is not sensitive to them may go badly astray. The first is an orientation to the turn-taking issue: who will talk next; the second is an orientation to the action implications for non-addressed parties of utterances designed for their addressees; the third is the issue of schism, i.e., the problem in extended discourse in multi-party interactions of maintaining a single discursive arena in the face of the potential for the breaking up of the interaction into two or more separate conversations.

First, who will talk next. In contrast to most other treatments of conversation, discourse or other formulations of talk-in-interaction which focus on "dialogue," conversation-analytic work has from the outset found it necessary to address data with varying numbers of participants. In part, this is because one of the underlying organizations of talk-in-interaction, the turn-taking organization by which opportunities to participate get distributed, cannot plausibly be taken to be differently designed for each discrete number of participants, and does not appear to follow some straightforward algorithm with increasing numbers. As my late colleague Harvey Sacks remarked early in his explorations of turn-taking, although two-party conversation appears to alternate formulaically—ABABAB, three-party conversation does not proceed ABCABCABC.

---

21 I say "some instances of discourse" because settings with more than two *persons* may nonetheless be self-organized for the purposes of talk-in-interaction into two parties (cf. Sacks, Schegloff and Jefferson, 1974; Schegloff, 1995).

Rather, turn-taking appears to be formally organized, that is, for any number of parties rather than particular numbers, and there appear to be orderly practices by which opportunities to speak are allocated among the parties in ways which also constrain the size of what can be done in those opportunities. The apparently formulaic alternation in the case of two-party interaction is, then, a special case of a more general and formal type of organization.[22] In conversation that allocation is administered by the participants locally, that is, allocation of next turn is the product of practices implemented in the current turn.

The consequence is that whatever else some speaker may be doing in a turn, whatever information may be distributed in it or whatever actions may be done through it, one issue systematically relevant in it, for which interlocutors parse it, is its bearing on the allocation of next turn—an issue which becomes organizationally consequential with more than two parties, when formulaic alternation gives way to contingent distribution. Because this issue infiltrates and permeates the talk rather than constituting separately articulated expressions in it (the practices of addressing aside), it can enter into the shaping of various aspects of discourse, and an increasingly sophisticated discourse analysis will have to be attentive to it.

For example, the selection of reference terms for persons (Sacks and Schegloff, 1979) or for places (Schegloff, 1972) is highly sensitive to considerations of "recipient-design;" that is, speakers are ordinarily charged with using forms of reference which recognize and exploit what the targeted recipient of the utterance is supposed to know. By incorporating reference terms differentially accessible to different co-participants, an utterance can be constructed to select one co-participant in particular to respond next (cf. Goodwin, 1979, 1981); such a selection may at the same time make necessary supplementary identification of what/who has been thereby referred to for those to whom the reference form is not recognizable. Thus, both the composition and the extensiveness of the discourse may be shaped by considerations related to next speaker selection.

A second issue which can become salient in interaction involving more than two parties is an orientation to the action implications for non-addressed parties of utterances designed for their addressees. Here I will be brief, for this issue is not in principle systematic, as the prior one was, but is occasional. Some utterances, by virtue of the action which they implement vis-à-vis one interlocutor, can be understood as doing another, related action to a different interlocutor (I described it once as a "derived action," Schegloff, 1984). If I compliment a contributor to this volume as the most elegant writer on discourse, I risk insulting others who feel slighted thereby.[23] Although it turns out that this potential can be realized in two-

---

[22] More generally, two-party talk-in-interaction can be the site of organizational problems and practices (both ones bearing on turn-taking issues and ones relevant to quite disparate themes and exigencies) whose major provenance is multi-party interaction, and which are simply inherited into an environment which does not especially give rise to them.

[23] Sacks developed a penetrating analysis of "safe compliments" be reference to their avoidance of this vulnerability (Sacks, 1992: I:60–61, 464–465, 597 ff., et passim).

party interaction as well (Schegloff, 1984), it is more exposed as a problem in multi-party interaction, and can impact on the construction and realization of discourse.

The third issue I want to mention is the issue of schism, i.e., maintaining a single interactional arena vs. the breaking up of the interaction into two or more separate conversations. This is again related to the number of participants. The issue of who will be next speaker emerges when there are three or more participants. The possibility of schism emerges when there are four or more. Its bearing on discourse—and especially on extended talk articulated by a single speaker—is that, as long as the extended speaker retains a single hearer, other parties to the interaction may gradually disengage, and, beginning with collusive side exchanges, may gradually develop a separate conversation of their own. The most effective deterrent to this development is the distribution of opportunities to talk among the several participants—including at times the forced draft of ones who appear in danger of drifting into schisms.[24] Because discourse (in the sense of multi-sentential productions) in particular is vulnerable to this dynamic, discourse design regularly is sensitive to it, and the production of what could have been extended discourses as shorter bursts of talk may often be understood in part by reference to this and related organizational issues. Analysts of discourse therefore need to be alert to this area.

5. Contingency

One last note in closing. The several themes to which I have called attention all involve a major challenge to computational interests in discourse, though they are hardly alone in posing this challenge. The challenge is *contingency*. Although the organization of talk-in-interaction is orderly (else it would be opaque to its participants), it is characterized by contingency at virtually every point.

The action(s) which some utterance implements is often a contingent product of its interactional setting. Although the orderly basis for various possible understandings of an utterance can be explicated, which understandings (e.g., of the action it is doing) will actually be entertained on a particular occasion may not predictable. And, as we have seen, the action which some utterance component enacts can turn out to be retroactively contingent; having implemented an announcement and complaint when brought to first possible completion, subsequent increments can recast it as the start of a different action (without "falsifying" the initial understanding of it, either by interactional co-participant or by professional analyst).

In implementing some action(s), an utterance can make a range of sequelae or responses contingently relevant next. Which of alternative contingent next actions a next speaker will do, however, is not in principle predictable. Still, although

---

24 See, for example, Goodwin, 1987a for analysis of such an episode. On the full development of schisms and their subsequent resolution, cf. Egbert, 1993.

whether an invitation will be accepted or declined, for example, is in principle indeterminate, much can be said about how either will be done if it is chosen—for example, whether it will be done promptly or delayed, explicitly or indirectly, baldly or with an account, etc. To be sure, that is also contingent, but there are orderly types of inferences which are observably generated if that type of next action is not done in that way—if, for example, an invitation is rejected precipitously, directly, explicitly and with no account. The co-participation of interlocutors in the production of talk, though a principled feature of talk-in-interaction, is always contingent in its occasioned expression. There are various places at which another can initiate talk and action, various practices for doing so, and (in multi-party interaction) alternative participants who can do so. But who, when and where are always contingent. There is virtually nothing in talk-in-interaction which can get done unilaterally, and virtually nothing which is thoroughly pre-scripted.

Contingency—interactional contingency—is not a blemish on the smooth surface of discourse, or of talk-in-interaction more generally. It is endemic to it. It is its glory. It is what allows talk-in-interaction the flexibility and the robustness to serve as the enabling mechanism for the institutions of social life. Talk-in-interaction is permeable; it is open to occupation by whatever linguistic, cultural, or social context it is activated in. It can serve as the vehicle for whatever concerns are brought to it by the parties engaging it at any given time. One underlying "burning" issue for computational interests in discourse analysis is how to come to terms with the full range of contingency which talk-in-interaction allows and channels. The themes of action, interaction and multi-party interaction on which I have focussed are three strategic—and I suspect under-appreciated—loci of this contingency.

Postscript

A referee of an earlier draft of this paper concluded a graciously appreciative assessment with a juxtaposition of its "whole method of analysis" to the referee's own "reservations," ones thought "likely to be shared by other readers," and suggested the possibility that "the author may want to explicitly address them in a preamble." I welcome the suggestion, though the reader will have noticed that I have preferred a post-amble, as it were, though I disavow the air of leisureliness which the neologism may hint at. If a reader shares the reservations, they will have been prompted by the paper, and should be addressed after that prompting, not before it. So here are the referee's reservations. I give them en bloc, and then take them up one by one.

> The problem for me is that the approach is purely descriptive and the analysis seems post hoc. The account of the data given by Schegloff seems persuasive, but is there any way of checking its validity? It seems that Schegloff is really setting up a hypothesis—(or really a set of hypotheses) that is interesting and plausible but which remains untested. My concern is

that someone else could come along and offer a very different interpretation of a conversation. (To some extent, we had examples of this in the discussion at the Burning Issues meeting, where there were alternative accounts suggested for the first conversation). If so, how would we know which interpretation to choose? Do we just rely on the intuitions of native conversationalists, or can we put this kind of hypothesis to more stringent test? In short, I'd like to see the insights that he has gained about what seem to be regularities in how conversations are managed firmed up into much more specific statements that would be objectively testable. Take, for instance, the interpretation of the Donny/Marcia conversation. As one with experimental leanings, I'd like to see someone do a study where conversations were contrasted in which A is trying to persuade a reluctant B to do something, as opposed to a situation where A is requesting help from a willing B. (Even if one didn't want to set up an artificial situation, it would be possible to use real data, by getting naive listeners to classify conversations as one kind or the other and then contrasting the transcripts of conversations that represented these two kinds.) The prediction would be that B would be silent more often in the first type of situation. I'm not suggesting that Schegloff should rush off and run such experiments, but I'd be interested to know whether he thinks this would be a valid next step—or if not, why not? He is critical of those who adopt more formal modelling approaches that they do not incorporate features such as interaction and timing into their analyses; I think that they would be very ready to do so if he could state much more explicitly the regularities that one sees and their significance. So it is crucial to know whether he thinks that this is in principle possible, even if it is not a currently achievable goal. If he doesn't, then there seems little hope of any integration between those adopting more computational approaches and researchers such as Schegloff.

A variety of issues are presented here, only some of which can be taken up, and those only in a truncated (if not peremptory) fashion.

1) The problem for me is that the approach is purely descriptive and the analysis seems post hoc. The account of the data given by Schegloff seems persuasive, but is there any way of checking its validity? It seems that Schegloff is really setting up a hypothesis—(or really a set of hypotheses) that is interesting and plausible but which remains untested.

1a) The analysis is surely post hoc, in the plain sense that it is done after the events being examined. So is most of astronomy, geology, paleontology, etc. Response to the characterization of the work as "purely descriptive" depends on what that is taken to mean, and what it is taken to contrast with. If it is taken to contrast with experimental studies designed around explicitly causal hypotheses or "formal models," then it is plainly the case; as there is no element at all of experiment, explicit causal hypotheses, or formal models here (though conversation-analytic work of the sort presented in Sacks et al., 1974 or Schegloff et al., 1977 is criticized by some for its formalism), it is "purely descriptive." If that is meant to deny, however, that there is an account of how the conversational episodes examined came to have the actual, specific, detailed trajectory which they did, then it seems on the face of it incorrect. The analysis offered here is full of claims about how various occurrences in the talk were heard and understood, how subsequent conduct by the other party gives evidence that that is indeed how they understood it, proposes how the next bit of conduct is to be understood as responsive to what preceded and as relative to alternative sequelae which previous research has shown to be alternative possibilities, and how the production of that alternative moves the interaction down a particular path or trajectory and/or embodies some already described formal organization. If accounts of how something comes to be—in detail—how it is are understood as involving some element of explanation, then the approach taken here is *not* "purely descriptive."

1b) If the account seems "persuasive," then one might look to that account for a first check on its validity—for part of its persuasiveness is that those who find it persuasive take it to formulate with some adequacy the actual processes at work in the data, and that is what validity is most directly about. Persuaded readers might then begin by asking themselves what makes it persuasive for them. But I am not abandoning them; there are ways of checking validity, and they (or some of them) are already in the paper. If it is "a hypothesis" that, for example, "Guess what" is a possible pre-announcement, then one of the directly relevant claims being entered is that it is taken by its recipient to be a possible pre-announcement (or could have been)—for it could hardly matter less that we analysts call it that or treat it as that if the interlocutors do not. What else could we mean by the claim anyway? There is a way of checking the validity of this claim, and that is to examine what the recipient of this utterance does in its immediate aftermath for some display of how they understood what preceded. (That it is to the initial aftermath that we should look in the first instance is a very general, multiply documented finding in a broad and extensive range of conversation-analytic research.) As the foregoing text has already presented a brief version of this analysis, and a number of other such analyses, I will not repeat it/them here. I simply want to invite readers to re-frame their understanding of these earlier discussions as precisely seeking and explicating evidence that the accounts being offered of/for each element of the interaction are "valid"—i.e., represent the understanding of the participants of that element and how it figures in what is going on in the setting.

> 2) My concern is that someone else could come along and offer a very different interpretation of a conversation. (To some extent, we had examples of this in the discussion at the Burning Issues meeting, where there were alternative accounts suggested for the first conversation). If so, how would we know which interpretation to choose? Do we just rely on the intuitions of native conversationalists, or can we put this kind of hypothesis to more stringent test? In short, I'd like to see the insights that he has gained about what seem to be regularities in how conversations are managed firmed up into much more specific statements that would be objectively testable.

2a) The possibility of "different interpretations" is a common theme in reactions to work of the sort presented here, perhaps because the data with which it deals appear to be *prima facie* accessible to vernacular understanding and interpretation. Indeed they *are* vernacularly accessible, for one job which a society's culture does for its members is to provide the resources for the "common sense" or "practical" analysis of what goes on in interaction. That does not mean, however, that any old interpretation will do. Or, indeed, that it is trivially easy to provide alternative interpretations in the first place. Or ones at a comparable level of detail—or *any* level of detail. Or ones for which supportive accounts can be given of the sort just discussed under point 1b—i.e., evidence that the new interpretation is grounded in the *demonstrable* orientations of the co-participants *in the interaction*, as evidenced in their observable conduct.

But if someone does offer another interpretation, or analysis, and it does address observable details of the interaction being examined, and it can be grounded in the activities of the participants as displayed in their conduct, then what is the problem? It may well elaborate, enrich, laminate, complement, etc., the analysis which I have offered (or some other analyst has offered in some other inquiry). Or it may be arguably (or even demonstrably) incorrect. For example, one account offered of the Debbie/Nick conversation at the meeting was that Debbie was "coming on" to Nick, i.e., was engaged in a form of seductive behavior. If the claim is that she initiated the conversation to do this, then I believe that it is demonstrably incorrect, though there is not the space here for me to explicate the demonstration.[25] If the claim is that there are elements of "courtship-related" talk informing Debbie's participation in the conversation, then I am inclined to agree, although I would want to treat these together with the ways in which Nick's talk is produced, and this might qualify the somewhat one-sided attribution of "coming on" to Debbie. However, it is unclear what bearing this "different interpretation" has on the analysis offered earlier in this paper. One potential bearing—which relates Debbie's questioning uptake of Nick's claim to have his waterbed to his persistent teasing earlier in the conversation—is noted in the earlier analysis, but is tied to Nick's teasing without relating that teasing to courtship ritual.

In sum, it is non-trivial to provide different interpretations which have a prima facie claim to be taken seriously, but if such alternative accounts are offered, there are ways of evaluating them—the same ways employed in grounding the account offered in the first instance. If alternative accounts do well under such examination, then they may be compatible with prior analysis, in which case we have a net gain—an enrichment of the analysis (as in the relationship of Schegloff, 1992d to Goodwin, 1987b, whose data it re-examines). Or if they are not compatible, then we have to figure out some way to choose—just as we do in any other systematic, disciplined form of empirical research.[26]

---

25 It involves evidence at both the outset and the closing of the conversation that Debbie called in the first instance to talk to her boyfriend, Mark, and not to Nick.

26 Indeed, it could be argued that in a great many other research paradigms and programs the point of direct contact between analyst and data is hidden from scrutiny in coding operations, technical equipment registering only outcome/measures, etc. which insulates the primary analysis on which all subsequent analysis is built from "different interpretations;" all potential variation and discrepancies in the relationship between data and analysis is "managed" by waving the magic wand of conjectured randomization of error. In the procedures followed in the present analysis, the reader is shown the primary data and the primary analysis, as well as warrants for that analysis. Why this procedure should engender

2b) That said, and quibbles over terminology aside, it may be useful to insist on a distinction between "interpretation" and "analysis," because the terms carry with them virtually ineradicable traces of their vernacular usages—in which "interpretation" is essentially contestable and invites alternatives no matter how compelling, whereas "analysis" carries the possibility of definitiveness (a definitiveness which is not incompatible with alternatives, but does not imply them). One key difference between "interpretation" and "analysis" in this domain of inquiry (as I understand it) is that analysis lays bare how the interpretation comes to be what it is—i.e., what about the target (utterance, gesture, intonation, posture, etc.) provides for the interpretation which has been proposed for it. So interpretation may be more or less subtle, deep, insightful, etc., but remains vernacular interpretation nonetheless; the issue is not its excellence. Analysis is "technical;" it explicates by what technique or practice the interpreted object was composed and produced, and by what technique or practice of uptake the interpretation was arrived at. And analysis grounds those claims in the observable conduct of the parties whose interaction is being examined.

2c) In any case, nowhere in the preceding analysis is it proposed or implied that we "just rely on the intuitions of native conversationalists." The analyses offered above have undoubtedly been informed by my intuitions as a native conversationalist, but it is not in that capacity that I ask colleagues to take them seriously. They are offered as technical analysis. And it is with other technical analyses that they are to be juxtaposed. What I have said in the prior two points concerns precisely the issue of making competing accounts into technical analyses. Curiously, it is those experimentally inclined investigators who wish to put interactional materials before naive judges and treat their reactions seriously (see below) who seem to me to wish to "just rely on the intuitions of native conversationalists." Indeed, I submit that the accounts offered above have been put "to a more stringent test," and have been "firmed up into much more specific statements that [have been] objectively test[ed]."

3) Take, for instance, the interpretation of the Donny/Marcia conversation. As one with experimental leanings, I'd like to see someone do a study where conversations were contrasted in which A is trying to persuade a reluctant B to do something, as opposed to a situation where A is requesting help from a willing B. (Even if one didn't want to set up an artificial situation, it would be possible to use real data, by getting naive listeners to classify conversations as one kind or the other and then contrasting the transcripts of conversations that represented these two kinds.) The prediction would be that B would be silent more often in the first type of situation.

3a) Although it is tempting, I will forego the opportunity to discuss the problems with setting up "artificial situations" for studying interaction along the lines embodied in the present paper. Suffice it for now to say that proceeding in that

---

greater concern about "different interpretations" may be understandable, but is not clearly justifiable.

fashion would quickly involve constraints such as "holding everything constant" in a domain where we are still finding out what "everything" should be taken to include, in which it does not appear possible to "control" what we already know to be included, and in which there is no good reason to think that "it all evens out." So I will stick to the proposal "to use real data." It is a touch ironic to point out that no loss in rigor (or "scientificity") is necessarily entailed in opting for "real data." There is, after all, an alternative "scientific" rhetoric (and "paradigm") often termed naturalistic, exemplified in disciplines such as astronomy, ethology, geology, paleontology, etc. Indeed, in many areas successful experimentation has followed a stage of naturalistic inquiry in which the parameters of the target domain were established by observation, as were the terms and conditions for viable experimental inquiry.

Although the analytic practice exemplified here has not availed itself of "naive listeners" (although our data sources have undoubtedly included some!), there have in fact been systematic efforts to compare the sorts of sequences in question here, including the ways in which silence figures in them. In fact, there is a robust literature in this area (which goes under the name of studies of preference/dispreference). Some of the relevant findings are reviewed in Atkinson and Drew, 1979: Chapter 2; Heritage, 1984b:265–292, Levinson, 1983:332–356, Schegloff, 1988d:442–457; among the relevant papers reporting these findings are Davidson, 1984, 1990; Drew, 1984; Pomerantz, 1978, 1984; Sacks, 1987[1973]; and Schegloff, Jefferson and Sacks, 1977.

"The prediction," by the way, "that B would be silent more often in the first type of situation" is, generally speaking, on target. However, researchers working along conversation-analytic lines prefer (if I may again quote the referee) "...to see the insights...about what seem to be regularities in how conversations are managed firmed up into much more specific statements that would be objectively testable." It has seemed to us that for an organization of interaction which is always employed by participants in singular occasions and moments, the relevant orderliness in the deployment of silence would be underspecified if characterized only as "more often in the first type of situation." The conversation-analytic literature, accordingly, is rather more specific about where the silences are relative to the structure of turns, relative to the types of actions being done in a turn, relative to the structure of sequences, relative to other forms of conduct which may enhance or mitigate or qualify the import of the silence, etc.[27]

> 4)   I'm not suggesting that Schegloff should rush off and run such experiments, but I'd be interested to know whether he thinks this would be a valid next step—or if not, why not? He is critical of those who adopt more formal modelling approaches that they do not incorporate features such as interaction and timing into their analyses; I think that they would be very ready to do so if he could state much more explicitly the regularities that

---

[27] The findings in question here were all derived from studies of naturally occurring interaction. Whether that qualifies under the criterion of "objectively testable" I do not know. If the issue regarding objectively testable is quantification, cf. the discussion in Schegloff, 1993.

one sees and their significance. So it is crucial to know whether he thinks that this is in principle possible, even if it is not a currently achievable goal. If he doesn't, then there seems little hope of any integration between those adopting more computational approaches and researchers such as Schegloff.

4a) I do hope that students of discourse will come to appreciate that criticizing "formal modelling approaches" for not incorporating "features such as interaction and timing into their analyses" is not a stylistic or paradigmatic or political option which one may or may not adopt. It is like not incorporating gravity and electricity into one's model of the physical world. They are, so far as we can tell, naturally occurring, indigenous properties of co-present social interaction and many of its transformations (e.g., talk on the telephone), and ones which are thus "present at the birth" of discourse in what seems its primordial provenance. "Interaction and timing" are not simply two more variables to be added in after other, supposedly more basic, factors—such as propositional content, information structure, syntactic organization, lexical composition, semantic specification, phonological realization, prosodic shaping, and articulatory enactment — have done their work. All of those "factors" do their work within a situation fundamentally shaped by—no, *constituted* by—interactional considerations, structures and constraints and in an ineluctably temporal world, whose temporality (as it happens) has been made organizationally relevant to the way interaction works. So it is not a matter of taste whether to incorporate them or not. It is a matter of dealing with the world as it is, as best we now understand it.

As suggested in the discussion in point 3a above, there is a larger literature available in this area than many students of discourse seem familiar with. Perhaps if serious researchers examined the best of that literature seriously, and thought about how to incorporate it in their own work, they might find it sufficiently explicit and specific to be of use. If not, perhaps they could tell those of us who study talk-in-interaction naturalistically where the problems are which trouble them, and we could try to be helpful within the canons of rigorous work as we understand them from having tried to think hard about our materials.

I am not optimistic about the use of experiments which compromise the naturally occurring constitution of talk-in-interaction for reasons which I hinted at earlier, and whose elaboration is not possible here. This is not a principled objection to experimentalism per se, but to the at present non-calculable effects of imposed artifice on the conduct of interaction. But elements of experimental inquiry can be combined with naturalistic inquiry in ways which do not compromise the naturalistic integrity of the empirical materials, and such undertakings may appeal to some who sympathize with the referee's comments. I end with a case in point, a project deveoped several years ago in collaboration with a neurolinguist—Dr. Diana Van Lancker, though not in the end carried through.

My colleague had worked within an experimentalist paradigm on problems experienced by persons who had suffered trauma to the right hemisphere of the brain with the recognition of familiar voices. A mutual colleague had told me that this was the topic of her research and we got together because I too had worked on the recognition of familiar voices (Schegloff, 1979b). It turned out that the voices she was concerned with were those of Kennedy, Churchill, Bob Hope, etc., as presented in brief taped excerpts from public occasions, presented under experimentally controlled conditions to experimentally partitioned sub-populations. The familiar voices with which I had been concerned were those of one's spouse, or parent, or child, or close friend, or work associate, as presented at the start of a telephone conversation to one expected to recognize the speaker from a very small voice sample (typically only "Hi") in the course of ordinary, mundane conversations. There was here a marvelous opportunity to combine naturalistic with experimental research. We planned to secure permission to tape record the bedside telephones of recent victims of right hemisphere brain insults, and hear how they dealt with the first moments of calls in which friends' and intimates' voices would be presented for possible recognition. The patients for the study could be selected according to any experimental protocol that seemed desirable; my collaborator could do the formal testing using snippets of tape from famous people supporting that research program. I had no objection to the experimental framework for this research for it left uncompromised the naturalistic auspices of the data with which my analysis would have to come to terms. We could then compare recognition of familiar/intimate voices with recognition of familiar/celebrity voices, recognition in experimental test situations with recognition as part of a common interactional context of the society, and begin to explicate the ways in which our understanding of brain function could be specified, and our understanding of the artifacts of experimentation in this area illumined.[28]

It does not seem to me that the future lies in the direction of that kind of experimentation (limited as it is) which can be made compatible with serious disciplined work on naturally occurring interaction. But perhaps the perceived necessity of such experimental work is an artifact of methodological traditions whose serious relevance is waning. If those who favor computational approaches to discourse find the sort of work presented here of potential interest, the next step may best be not denaturing it by trying to graft it onto experimental formats, but rather seriously pursuing it in its own terms, trying to understand why researchers proceed as they do in this area, and then thinking through what changes this work might suggest for how computationally oriented work is done, rather than how conversation-analytic work might adapt to them.

Successful convergence here is, after all, a long shot. The problem of contingency with which my paper began poses truly formidable obstacles to computational approaches. But if some useful interchange between these modalities of work is to be realized, it is most likely to come not from transforming the object from which you would like to learn, but from taking it

---

[28] Other such efforts to combine elements of experimentalism and naturalism in ways which avoid—or minimize—compromising the integrity of the data include some of the work of Herbert Clark and his associates, e.g., Clark, 1979 or Clark and French, 1981.

seriously in its own terms. In the end, it will be the computationalists who will have to figure out how to do this. I hope we will be allowed to help.

Note

Parts of this paper have previously appeared under the title "Discourse as an Interactional Achievement III: On the Omnirelevance of Action" in the journal *Research on Language and Social Interaction*, by permission of the editors and publisher. That title alludes to two earlier papers on the theme "Discourse as an Interactional Achievement" (Schegloff, 1981, 1987, 1988), which are relevant here as well. My thanks to John Heritage, Sally Jacoby, Sandra Thompson and the editors and referees for helpful comments on earlier drafts of the present effort.

References

Atkinson, J.M. and Drew, P. (eds.) (1979). Order in Court: The Organization of Verbal Interaction in Judicial Settings. London: Macmillan.

Clark, H.H. (1979). Responding to Indirect Speech Acts. Cognitive Psychology 11(4).

Clark, H.H. and French, J.W. (1981). Telephone 'Goodbyes.' Language in Society 10:1–19.

Davidson, J.A. (1984). Subsequent Versions of Invitations, Offers, Requests, and Proposals Dealing with Potential or Actual Rejection. In J.M. Atkinson and J.C. Heritage (eds.), Structures of Social Action. Cambridge: Cambridge University Press. 102–128.

Davidson, J.A. (1990). Modifications of Invitations, Offers and Rejections. In George Psathas (ed.), Interaction Competence. Washington, D.C.: International Institute for Ethnomethodology and Conversation Analysis/ University Press of America. 149–180.

Drew, P. (1984). Speakers' Reportings in Invitation Sequences. In J.M. Atkinson and J.C. Heritage (eds.), Structures of Social Action. Cambridge: Cambridge University Press. 152–164.

Duranti, A. and Brenneis, D. (eds.) (1986). Special Issue on the Audience as Co-Author. Text 6(3).

Egbert, M. (1993). Schisming: The Transformation from a Single Conversation to Multiple Conversations. Unpublished Ph.D. Dissertation, Department of Applied Linguistics, University of California, Los Angeles.

Erickson, F. (1992). They Know All the Lines: Rhythmic Organization and Contextualization in a Conversational Listing Routine. In P. Auer and A. di Luzio (eds.), The Contextualization of Language. Amsterdam: John Benjamins. 365–397.

Ford, C. (1993). Grammar in Interaction: Adverbial Clauses in American English. Cambridge: Cambridge University Press.

Ford, C. and Thompson, S.A. (forthcoming). Interactional Units in Conversation: Syntactic, Intonational and Pragmatic Resources for the Management of Turns. In E. Ochs, E.A. Schegloff and S.A. Thompson (eds.), Interaction and Grammar. Cambridge: Cambridge University Press.

Goodwin, C. (1979). The Interactive Construction of a Sentence in Natural Conversation. In G. Psathas (ed.), Everyday Language: Studies in Ethnomethodology. New York: Irvington Publishers. 97–121.

Goodwin, C. (1981). Conversational Organization. New York: Academic Press.

Goodwin, C. (1987a). Forgetfulness as an Interactive Resource. Social Psychology Quarterly 50(2): 115–130.

Goodwin, C. (1987b). Unilateral Departure. In Graham Button and John R.E. Lee (eds.), Talk and Social Organization. Clevedon, England: Multilingual Matters. 206–216.

Goodwin, M. H. (1980). Processes of Mutual Monitoring Implicated in the Production of Description Sequences. Sociological Inquiry 50: 303–317.

Heritage, J. (1984a). A Change-of-State Token and Aspects of Its Sequential Placement. In J.M. Atkinson and J.C. Heritage (eds.) Structures of Social Action. Cambridge: Cambridge University Press. 299–345.

Heritage, J. (1984b). Garfinkel and Ethnomethodology. Oxford: Polity Press.

Jefferson, G. (1988). On the Sequential Organization of Troubles-Talk in Ordinary Conversation. Social Problems 35(4): 418–441.

Jefferson, G. and Lee, J.R.L. (1981). The Rejection of Advice: Managing the Problematic Convergence of a 'Troubles-Telling' and a 'Service Encounter'. Journal of Pragmatics 5: 399–422.

Jefferson, G., and J. Schenkein (1978). Some Sequential Negotiations in Conversation: Unexpanded and Expanded Versions of Projected Action Sequences. In J. Schenkein (ed.), Studies in the Organization of Conversational Interaction. New York: Academic Press. 155–172.

Labov, W. and Fanshel, D. (1977). Therapeutic Discourse. New York: Academic Press.

Lerner, G. H. (1987). Collaborative Turn Sequences: Sentence Constructionand Social Action. Unpublished Ph.D. dissertation, School of Social Science, University of California, Irvine.

Lerner, G. H. (1991). On the Syntax of Sentences-in-Progress. Language in Society 20: 441-458.

Lerner, G. H. (forthcoming). On the 'Semi-Permeable' Character of Grammatical Units in Conversation: Conditional Entry into the Turn Space of Another Speaker. In E. Ochs, E. A. Schegloff and S.A. Thompson (eds.) Interaction and Grammar. Cambridge: Cambridge University Press.

Levinson, S. (1983). Pragmatics. Cambridge: Cambridge University Press.

Local, J. (1992). Continuing and Restarting. In P. Auer and A. di Luzio (eds.), The Contextualization of Language. Amsterdam: John Benjamins. 273–296.

Mandelbaum, J. (1987). Couples Sharing Stories. Communication Quarterly 35(2):144–170.

Mandelbaum, J. (1989). Interpersonal Activities in Conversational Storytelling. Western Journal of Speech Communication 53(2): 114–126.

Pomerantz, A. (1978). Compliment Responses: Notes on the Co-operationof Multiple Constraints. In Jim Schenkein (ed.), Studies in the Organization of Conversational Interaction. New York: Academic Press. 79–112.

Pomerantz, A. (1984). Agreeing and Disagreeing with Assessments: Some Features of Preferred/Dispreferred Turn Shapes. In J.M. Atkinson and J.C. Heritage (eds.), Structures of Social Action. Cambridge: Cambridge University Press. 57–101.

Sacks, H. (1987 [1973]). On the Preferences for Agreement and Contiguity in Sequences in Conversation. In G. Button and J.R.E. Lee (eds.), Talk and Social Organization. Clevedon: Multilingual Matters. 54–69.

Sacks, H. (1992). Lectures on Conversation. Volumes 1 and 2. Edited by Gail Jefferson. Oxford: Basil Blackwell.

Sacks, H. and E.A. Schegloff (1979). Two Preference in the Organization of Reference to Persons in Conversation and Their Interaction. In G. Psathas (ed.), Everyday Language: Studies in Ethnomethodology. New York: Irvington Publishers. 15–21.

Sacks, H., E. A. Schegloff, and G. Jefferson (1974). A Simplest Systematics for the Organization of Turn-Taking for Conversation. Language 50: 696–735.

Schegloff, E.A. (1972). Notes on a Conversational Practice: Formulating Place. In D.N. Sudnow (ed.), Studies in Social Interaction. New York: Free Press/MacMillan. 75–119.

Schegloff, E. A. (1979a). The Relevance of Repair for Syntax-for-Conversation.   In T. Givón (ed.), Syntax and Semantics 12: Discourse and Syntax. New York, Academic Press. 261–288.

Schegloff, E. A. (1979b). Identification and Recognition in Telephone Openings. In G. Psathas (ed.), Everyday Language: Studies in Ethnomethodology.   New York: Irvington. 23–78.

Schegloff, E.A. (1980). Preliminaries to Preliminaries: 'Can I Ask You a Question?' Sociological Inquiry 50: 104–152.

Schegloff, E. A. (1982). Discourse as an Interactional Achievement: Some Uses of 'uh huh' and Other Things that Come Between Sentences. In D. Tannen (ed.), Georgetown University Round Table on Languages and Linguistics. Washington D.C., Georgetown University Press. 71–93.

Schegloff, E. A. (1984). On Some Questions and Ambiguities in Conversation. In J.M. Atkinson and J.C. Heritage (eds.), Structures of Social Action. Cambridge: Cambridge University Press. 28–52.

Schegloff, E. A. (1986). The Routine as Achievement. Human Studies 9: 111–151.

Schegloff, E. A. (1987). Analyzing Single Episodes of Interaction: An Exercise in Conversation Analysis. Social Psychology Quarterly 50(2): 101–114.

Schegloff, E. A. (1988a). Presequences and Indirection: Applying Speech Act Theory to Ordinary Conversation. Journal of Pragmatics 12: 55–62.

Schegloff, E.A.(1988b). Discourse as an Interactional Achievement II: An Exercise in Conversation Analysis. In D. Tannen (ed.), Linguistics in Context: Connecting Observation and Understanding. Norwood, NJ: Ablex.

Schegloff, E.A.(1988c). Goffman and the Analysis of Conversation. In P. Drew and T. Wootton (eds.), Erving Goffman: Exploring the Interaction Order. Cambridge: Polity Press. 89–135.

Schegloff, E.A. (1988d). On an Actual Virtual Servo-mechanism for Guessing Bad News: A Single Case Conjecture. Social Prioblems 35(4): 442–457.

Schegloff, E.A. (1989). Reflections on Language, Development, and the Interactional Character of Talk-in-Interaction. In M. Bornstein and J.S. Bruner (eds.), Interaction in Human Development. Hillsdale, NJ: Lawrence Erlbaum Associates. 139–153.

Schegloff, E.A. (1990). On the Organization of Sequences as a Source of 'Coherence' in Talk-in-Interaction. In B. Dorval (ed.), Conversational Organization and its Development. Norwood, NJ: Ablex. 51–77.

Schegloff, E.A.(1991). Reflections on Talk and Social Structure. In D. Boden and D.H. Zimmerman (eds.), Talk and Social Structure. Cambridge: Cambridge University Press. 44–70.

Schegloff, E.A.(1992a) To Searle on Conversation: A Note in Return. In John R. Searle et al. (eds.), (On) Searle on Conversation. Amsterdam and Philadelphia: John Benjamins. 113–128.

Schegloff, E.A.(1992b). Introduction, in Harvey Sacks: Lectures on Conversation, Volume 1. Edited by Gail Jefferson. Oxford: Basil Blackwell, ix–lxii.

Schegloff, E.A.(1992c). Repair After Next Turn: The Last Structurally Provided Defense of Intersubjectivity in Conversation. American Journal of Sociology 97(5): 1295–1345.

Schegloff, E.A.(1992d). In Another Context. In A. Duranti and C. Goodwin (eds.), Rethinking Context: Language as an Interactive Phenomenon. Cambridge: Cambridge University Press. 191–228.

Schegloff, E.A. (1993). Reflections on Quantification in the Study of Conversation. Research on Language and Social Interaction 26(1): 99–128.

Schegloff, E.A. (1996). Turn Organization: One Intersection of Grammar and Interaction. In E. Ochs, E. A.Schegloff and S.A. Thompson (eds.), Interaction and Grammar. Cambridge: Cambridge University Press. 52–133.

Schegloff, E.A. (1995). Parties and Talking Together: Two Ways in Which Numbers Are Significant for Talk-in-Interaction. In P. ten Have and G. Psathas (eds.), Situated Order: Studies in Social Organization and Embodied Activities. Washington, D.C.: University Press of America. 31–42.

Schegloff, E.A., Jefferson, G. and Sacks, H. (1977). The Preference for Self-Correction in the Organization of Repair in Conversation. Language 53: 361–382.

Terasaki, A. (1976). Pre-Announcement Sequences in Conversation.   Social Science Working Paper 99, School of Social Sciences, Irvine, California.

Uhmann, S. (1992). Contextualizing Relevance: On Some Forms and Functions of Speech Rate Changes in Everyday Conversation. In P. Auer and A. diLuzio (eds.), The Contextualization of Language. Amsterdam: John Benjamins. 297–336.

Zimmerman, D.H. (1984). Talk and Its Occasion: The Case of Calling the Police. In D. Schiffrin (ed.), Georgetown University Round Table on Languages and Linguistics 1984. Washington, D.C.: Georgetown University Press. 210–228.

Appendix A

<u>Debbie & Nick</u>

```
01                      ((Ring Ring))
02                      ((Click/Pick-up))
03      Nick:           H'llo
04      Debbie:         ·hh- 'z <Who's this,
05                      (0.2)
06      Debbie:         This'z Debbie
07                      (0.3)
08      Nick:           Who's this.
09      Debbie:         This'z Debbie
10      Nick:           This is >the Los Angeles Poli[ce<
11      Debbie:                                       [Nno:[((Laugh))
12      Nick:                                             [ha ha
                        [ha
13      Debbie:         [Hi Nicky how areya.
14      Nick:           O:kay
15      Debbie:         hh u- Did Mark go to Ohio?
16      Nick:           Ohio?
17      Debbie:         Uh huh¿
18      Nick:           I dunno did he?
19      Debbie:         ·hh I: dunn[o::]
20      Nick:                      [ ha]ha
21      Debbie:         Ny-
22      Nick:           Yeah I think he's (com-)/(still) (  )-
23                      when's Mark come back, Sunday?  ((off phone))
24                      (0.8)
25      Nick:           Yeah I think he's comin back Sunday=
26      Debbie:         =Tomorrow? Is Rich gonna go get 'im?
27                      (0.2)
28      Nick:           I guess
29      Debbie:         Or is he gonna ca:ll?
30                      (0.8)
31      Nick:           h! (h)I du(h)nno he didn't tell me=
32      Debbie:         =Oh:: you have nothin' t'do with it
33      Nick:           (n)ha ha
34      Debbie:         ·hhh Um:: u- guess what I've-(u-)wuz lookin'
```

```
35                    in the paper:-have you got your waterbed yet?
36    Nick:           Uh huh, it's really nice °too, I set it up
37    Debbie:         Oh rea:lly? ^Already?
38    Nick:           Mm hmm
39                    (0.5)
40    Debbie:         Are you kidding?
41    Nick:           No, well I ordered it last (week)/(spring)
42                    (0.5)
43    Debbie:         Oh- no but you h- you've got it already?
44    Nick:           Yeah h! hh=                      ((laughing))
45    Debbie:         =hhh [hh ·hh]                    ((laughing))
46    Nick:                [I just] said that
47    Debbie:         O::hh: hu[h, I couldn't be[lieve you c-
48    Nick:                    [Oh (°it's just) [It'll sink in
49                    'n two day[s fr'm now (then    )((laugh))]
50    Debbie:                   [      ((laugh))           ]
                      Oh no cuz I just got- I saw an ad in the
51                    paper for a real discount waterbed s'
52                    I w'z gonna tell you 'bout it=
53    Nick:           =No this is really, you (haven't seen)
54                    mine, you'll really like it.
55    Debbie:         Ya:h. It's on a frame and everythi[ng?
56    Nick:                                             [Yeah
57    Debbie:         ·hh Uh (is) a raised frame?
58    Nick:           °mm hmm
59    Debbie:         How: ni::ce, Whadja do with Mark's cou:ch,
60                    (0.5)
61    Nick:           P(h)ut it out in the cottage,
62                    (0.2)
63    Nick:           goddam thing weighed about two th(h)ousand
                      pound[s
64    Debbie:              [mn:Yea::h
65                    I'll be[:t
66    Nick:                  [ah
67                    (0.2)
68    Debbie:         Rea:lly
69                    (0.3)
70    Debbie:         ·hh O:kay,
71                    (·)
72    Debbie:         Well (0.8) mmtch! I guess I'll talk tuh Mark
                      later then.hh
73    Nick:           Yeah I guess yo[u will.[eh heh huh huh [huh
74    Debbie:                         [·hhh   [ W e : l l : - [eh
75                    heh ·hhthat that: (·) could be debatable too
                      I dunno
76                    (0.2)
77    Debbie:         Bu:t ·hh so um: ·hh=
78    Nick:           =So (h!) um [ uh [let's see my name's Debbie
                                                 [Idon't  ((laugh))
79    Debbie:                     [·hh [um      [  ((laugh))
80    Debbie:         ·hhh! Okay I'll see you later Nick=
81    Nick:           =Okay
82    Debbie:         Buh bye
83    Nick:           Bye bye
84                    ((phone hung up))
85                    ((click))
```

Selected transcription notational conventions

(Cf. Sacks, Schegloff and Jefferson, 1974; Atkinson and Heritage, 1984)

Um::	colons represent lengthening of the preceding sound; the more colons, the greater the lengthening.
I've-	a hyphen respresents the cut-off of the preceding sound, often by a stop.
^Already?	the circumflex represents sharp upward pitch shift; underlining represents stress, usually via volume; the more underlining, the greater the stress.
ni::ce	underlining directly followed by colon(s) indicates downward inflection on the vowel.
hhh hh ·hhh	represents aspiration, sometimes simply hearable breathing, sometimes laughter, etc.;
P(h)ut	when preceded by a superposed dot, it marks in-breath; in parentheses inside a word it represent laugh infiltration.
hhh[hh ·hh]	left brackets represent point of overlap onset;
[I just]	right brackets represent point of overlap resolution.
.,?	punctuation marks intonation, not grammar; period, comma and "question mark" indicate downward, "continuative," and upward contours, respectively.
( )	single parentheses mark problematic or uncertain hearings; two parentheses separated by an oblique represent alternative hearings.
(( ))	double parentheses mark transcriber's descriptions, rather than transcriptions.
(0.2)(·)	numbers in parentheses represent silence in tenths of a second; a dot in parentheses represents a micro-pause, less than two tenths of a second.
°mm hmm	the degree sign marks significantly lowered volume.

Perspectives from Linguistics

Chapter 2
Types of Structure: Deconstructing Notions of Constituency in Clause and Text

James R. Martin
University of Sydney

Clause and Text

In this paper I will present arguments in favour of a view of text structure in which constituency is not privileged, but deconstructed as just one way of looking at text organisation. This view of text structure has been developed in Australia in dialogue with Halliday's (e.g., 1994) and Matthiessen's (e.g., in press) work on English clause grammar. Consequently I will begin with an overview of their clause analysis before moving on to argue the main point of my paper—namely that constituency is a semantically biassed and reductive form of representation for text structure (i.e. that a text is not a tree).

1. Modes of Meaning at the Clause Level

In Systemic Functional Linguistics (hereafter SFL) interpretations of semiotic systems are organised with respect to metafunctions—highly generalised semantic components which shape paradigmatic and syntagmatic relations. Halliday (e.g., 1974, 1978, 1985) refers to these metafunctions as the ideational (including logical and experiential subcomponents), the interpersonal and the textual. Ideational resources construe experience as if it was natural reality; interpersonal resources construe social relations as intersubjective reality; and textual resources organise text/process (Martin 1985)—the semiotic reality which comes into being by way of construing ideational and interpersonal meaning. In SFL, this intrinsic functionality (Martin 1991), is projected onto context in register analysis in the proportions ideational to field, interpersonal to tenor and textual to mode. This tripartite model of intrinsic and extrinsic language function is outlined in Table 1.

Halliday (1979a) suggests that metafunctions organise syntagmatic relations as well as paradigmatic ones, and associates different types of structure with ideational, interpersonal and textual meaning. In his view ideational meaning uses *particulate* structuring principles, interpersonal meaning uses *prosodic* principles, and textual meaning *periodic* ones.

Table 1. Metafunctions and orders of 'reality'

generalised semiotic function	metafunction (organisation of language; intrinsic functionality)	register (organisation of context; projected extrinsic functionality)
Language for construing the social as intersubjective reality	interpersonal meaning	tenor
Language for construing experience as if 'natural' reality	ideational meaning	field
Language for organising text/process as semiotic reality	textual meaning	mode

Particulate structures are segmental. Experientially they divide bounded wholes into parts (as in constituency representation); logically they relate part to parts in potentially unbounded series (as in dependency representation). Prosodic structures are suprasegmental; they map over a range of segments, as with intonation and long components in phonology (cf. Palmer 1970, especially Waterson). Periodic structures are wave-like; they establish rhythmic peaks of prominence that bound units, as with Consonant Vowel Consonant, salient/nonsalient syllable, or tonic/nontonic foot alternations in phonology (Halliday 1967, 1985a). These correlations are summarised in Figure 1.

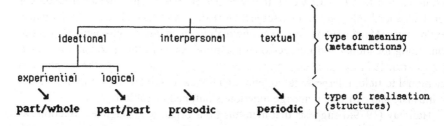

Figure 1. Types of meaning (metafunctions: at top) and types of structure (realisations: at bottom), after (Halliday 1979a) and (Matthiessen 1988)

Halliday is of course not alone in suggesting that constituency and dependency theory provide a very limited view of structuring principles. In articulating his principles, Halliday acknowledges the analogous perspectives recognised by Pike (1982) in tagmemic analysis. Pike's particle and wave correlate closely with

Halliday's conception, but field is closer to Halliday's system than his Firthian influenced notion of prosody. Rich interpretations of structure of this kind contrast most sharply with the formal syntax originating in or reacting to MIT research, where constituency representation is fundamental.

> Within tagmemic theory there is an assertion that at least three perspectives are utilized by Homo sapiens. On the one hand, he often acts as if he were cutting up sequences into chunks—into segments or *particles*...On the other hand, he often senses things as somehow flowing together as ripples on the tide, merging into one another in the form of a hierarchy of little *waves* of experiences on still bigger waves. These two perspectives, in turn, are supplemented by a third—the concept of *field* in which intersecting properties of experience cluster into bundles of simultaneous characteristics which together make up the patterns of his experience. (Pike 1982:12-13).

Matthiessen 1988 comments insightfully on the representational lag in SFL between the theory outlined above and the forms of representation which have evolved to implement the theory in language description. For example, Halliday (1985a) develops a distinct form of representation for logical meaning, clearly opposing interdependency to constituency (and thus logical construals of experience to experiential ones). But distinctive representations are not developed for prosodic and periodic patterns, which are expressed rather in constituent terms. In Halliday's (1981b,c) terms, logical part/part relations are expressed as univariate structures (structures emplying a single iterated variable), while experiential part/whole structures, interpersonal prosodic structures and textual wave structures are all represented in multivariate terms (i.e., using a fixed number of distinct variables). This representational problem is outlined in Figure 2.

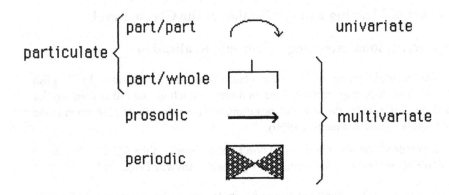

Figure 2. Types of structure and representational notations, from (Halliday 1985a)

By way of illustrating this representational strategy, consider Figure 3, which displays Halliday's multi-tiered analysis of an English clause (for univariate structure see Section 2.2.2). The same particulate form of representation is used for experiential meaning (Value Process Token), interpersonal meaning (Mood Residue) and for textual meaning (Theme Rheme). This faciliates mapping one tier onto another for purposes of text analysis and interpretation. But it does not really do justice to the prosodic impact of Mood over Residue (cf. Section 2.1), or the informational prominence of Theme over Rheme (cf. Section 2.3) in the English clause.

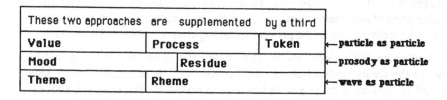

These two approaches	are	supplemented	by a third	
Value	Process		Token	←particle as particle
Mood	Residue			←prosody as particle
Theme	Rheme			←wave as particle

Figure 3. Halliday's multivariate renderings of particle, prosody and wave

It is suggested below that when analysing texts, oppositions among particulate, prosodic and periodic structure are just as significant as they are in clause analysis—and that accordingly, discourse models need to be developed which acknowledge these distinct structuring principles and provide forms of representation which accommodate them.

2. Types of Meaning and Realisation at the Clause Level

2.1 Interpersonal Meaning – Prosodic Realisation

A clear exemplification of prosodic clause structure is provided by English polarity. This is perhaps most striking in dialects which use *no* rather than *any* for indefinite deixis under the scope of negation; but the principle of realisation is the same across speakers (Fishman 1990).

> "If you don't get **no** publicity you don't get **no** people at the fight,"..."If you don't get **no** bums on seats you don't get paid...Anyway I enjoy it."

> (cf. standard *If you don't get any publicity for any fights in any papers from anyone ...*)

In examples such as these, negative polarity has been selected, established in the Mood element through the structural function Finite (*don't*), and then realised again across the Residue wherever indefinite deixis appears. As linguists have taken pains to point out to prescriptive school grammarians, negative clauses of this kind select once for polarity, and then map this meaning across the clause as opportunity presents itself. Thus the non-standard *no's* illustrated above do not cancel each other out; they simply reinforce the negative polarity ranging over the clause. Prosodic realisation of this kind in a sense naturalises the fact that polarity is a feature of the clause as a whole, not the particular clause segment which establishes it. Along these lines prosodic structure lends itself to interpersonal meanings (cf. McGregor 1990 on what he calls scopal relations). A representation for interpersonal structure of this kind is suggested in Figure 4.

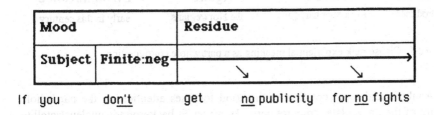

Figure 4. Interpersonal meaning realised as a prosody—polarity

Opportunistic realisation of this kind (cf. vowel harmony prosodies in Palmer 1970) represents one strategy deployed by languages for mapping prosodies onto experientially segmented stucture. An alternative strategy is to structurally demarcate the scope of interpersonal meaning through dependency structure. Tagalog (Martin 1990) makes use of this strategy to establish the domain of modal meanings, among others. This is illustrated below, where the linking particle *-ng* is used to construct the modal *sigurado* as head of the clause, with its domain dependent on it:

sigurado -ng   u-uwi   kang bahay ngayon   hapon

certain   lk[1]   go home   you-sg   house   today   afternoon

'You'll certainly go home to your house this afternoon.'

2.2 Ideational Meaning – Particulate Realisation

2.2.1 Experiential – Part/Whole or Nucleus/Satellite?

English transitivity provides a clear example of experiential construals of reality along particulate lines. Halliday (1985) proposes an ergative Agent-Process-Medium-Circumstance analysis for the activity realised in clauses such as *Early in*

---

[1] The 'lk' stands for linker, Tagalog's hypotaxis marker; see Martin (1995).

this century the Norwegians introduced explosive harpoons. An analysis of this kind takes the activity in question as a bounded whole, and divides it into four distinct parts, each playing a different role. The analysis is represented in constituent terms in Figure 5.

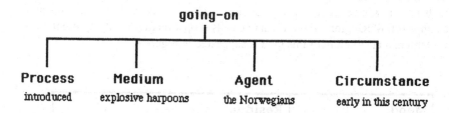

Figure 5. Clause rank experiential meaning as a part/whole configuration

Analysis and representation of this kind focusses attention on the part/whole nature of the particulate structure here. However, it backgrounds nucleus/satellite aspects of the construal. For example, as Halliday (1985a) reveals, the Process/Medium complex is fundamental to the description. It is the basis for the classification of processes into material, mental and relational classes; and the Medium is the one participant which regularly[2] appears without a preposition across process types. Agents on the other hand are optional and regularly appear with or without prepositions:

The Norwegians introduced explosive harpoons.
Explosive harpoons were introduced (by the Norwegians).

Circumstances are more peripheral still. Where absent, they are not necessarily implied (as with agentless passives); they regularly appear with a preposition; and they cannot be realised as Subject in effective[3] clauses (*This century was introduced explosive harpoons by the Norwegians*). In an alternative form of particulate representation, taking these phenomena into account, appears in Figure 6. Here the Process/Medium configuration is constructed as nucleus of the clause, with the Agent as an inner satellite and the Circumstance in outer orbit.

---

[2] The only apparent exception to this principle in English relates to the substitution of dispositive processes: *What did the Norwegians do to/do with the harpoons?*

[3] Some circumstances do have a restricted Subject potential in middle clauses: *This path's never been run on.*

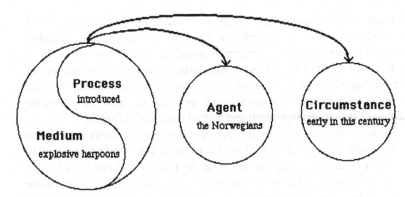

Figure 6. Clause rank experiential meaning as orbit: a nucleus with satellites

In Figure 7, an attempt is made to integrate the two perspectives, part/whole and nucleus/satellite. There, at clause rank, a constituency tree is deployed to relate parts to whole, and concentric ovals to capture perpherality patterns. The constituency perspective construes activity as a bounded whole and segments it; the orbital perspective focusses on a centre of activity, and then maps associated phenomena. It is suggested below that the orbital perspective is the one which can be most easily generalised across clause and text structure.

Part/whole construals of semiotic phenomena have widely deployed in 20th century linguistics. In grammar, clause segmentation of the kind discussed above have been regularly extended to lower ranks—through groups/phrases to words to morphemes. This limited extension of constituency is outlined in Figure 7, for the level of grammar.

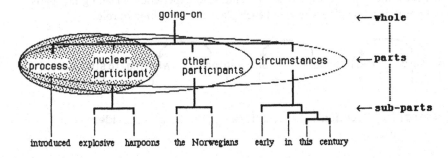

Figure 7. Nuclearity and constituency as two facets of transitivity structure (with extension of part-whole segmentation to lower ranks)

Early structuralist models in America pushed the metaphor even further, to describe the relationship between morphemes and phonemes. And linguists still write as if phonemes (and indeed lexemes) were composed of features; and as if texts were made up of clauses (e.g., Longacre 1976, 1979, Pike & Pike 1983). This constituency metaphor is urgently in need of deconstruction, including consideration of its bias towards experiential meaning, and of the influence of alphabetic forms of graphology which display texts as made up of paragraphs, which are made up of sentences, of words, of letters. The constituency card has almost certainly been dramatically overplayed, and that the price has been the marginalisation of logical, interpersonal and textual construals of reality.

Before turning to logical structure, it is perhaps important to clarify the sense in which the term embedding is deployed in this chapter. Here embedding will be used to refer to expansions of experiential meaning potential whereby a unit that has already been segmented reappears in decomposition. In *What the Norwegians did was introduce explosive harpoons,* for example, the clause is initially segmented into a Value-Process-Token structure. But instead of being filled by nominal groups, both Value and Token are realised by embedded clauses, which require a case segmentation of their own (Range-Actor-Process and Process-Goal respectively). Note that in SFL, embedding of this kind is distinguished from hypotaxis (cf. Section 2.2.2).

2.2.2 Logical Meaning—Part/Part or Multi-Nuclear?

English projection can be used to illustrate logical construals of reality in interdependency terms. The system is recursive, and verbal and mental processes project locutions and ideas respectively. Figure 8 contains an example of hypotactic projection, where a verbal process of saying projects a locution of thinking which projected an idea of wanting which projects the idea that whaling should stop. In structures of this kind, one segment gives rise to another, in an open ended interdependency series. In contrast to experiential meaning the 'parts' do not presume a bounded whole and each plays the same kind of role.

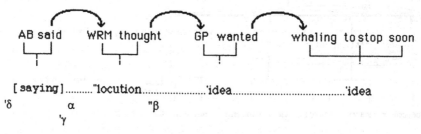

Figure 8. Logical meaning as interdependency in the context of projection

This structure can be further illustrated from Tagalog, a language which makes interdependency explicit through the hypotaxis marker-*ng/na* (appropriated by the language for prosodic purposes in the example below Figure 4; for discussion see Martin 1995). The structure for the meaning 'those naughty boys I saw yesterday' can be built up as follows, with *-ng* marking the dependence of one segment on another:

β	α
iyong	bata
that	child

γ	β	α
iyong	maraming	bata
		many

δ	γ	β	α
iyong	maraming	masamang	bata
			bad

δ	γ	β	α	β
iyong	maraming	masamang	batang	lalake
				man

δ	γ	β	α	β	γ
iyong	maraming	masamang	batang	lalakeng	nakita ko kahapon
					saw I yesterday

A structure of this kind has been developed regressively, leftwards from the head, *bata* 'child', and progressively to its right. Tagalog prefers deixis, numeration and description as premodifiers and classification and qualification as postmodifiers in nominal groups. A representation for this logical construal of meaning is offered in Figure 9.

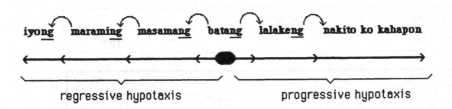

Figure 9. Nominal interdependency in Tagalog

Note that if an orbital perspective on experiential construals of reality is preferred over a part/whole one, then logical structures might be better referred to as serial rather than as part/part (the term *part* is a misnomer in any case for a structure not implying a whole). In these 'solar system' terms, the difference between experiential and logical structures is that experiential structures are mononuclear (i.e., one nuclus and one or more satellites) while logical structures are multi-nuclear (i.e., each satellite is itself a nucleus). From this point in the paper, particulate structures will be referred to as *orbital* (experiential) or *serial* (logical) along these lines.

The distinction between experiential and logical construals of experience has proven an important one in register analysis, particularly with respect to canonical differences between spoken and written discourse. Halliday (e.g., 1979b, 1985b) attributes part of the complexity of writing to experiential recursion (i.e. embedding as discussed above); this complexity is complemented in speaking by logical recursion (i.e., long series of interdependent clauses). Beaman (1984) and Biber (1988) accumulate evidence in favour of this distinction. Their research indicates that construing complexity simply in constituency terms under the label subordination provides a one-sided view of recursive structure that needs to be balanced by the interdependency perspective.

2.3 Textual Meaning—Periodic Realisation

English systems of theme and information exemplify the textual oganisation of semiotic reality into periodic patterns. Halliday (1985) suggests that first and last position in the English clause are constructed as complementary peaks of textual prominence. First position realises the function Theme, which specifies a text's orientation to its field (its angle on its subject matter); last position, where it is associated with the major pitch movement in the clause, realises the function New, which presents relatively newsworthy information from the field. These complementary peaks of textual prominence are outlined in Figure 10.

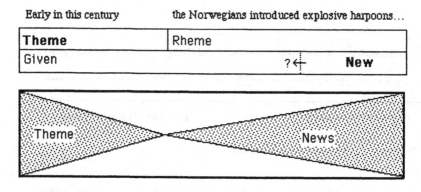

Figure 10. Clause rank textual meaning as a wave (pulses of prominence)

By definition, textual functions like Theme and New have no meaning apart from the role they play in contextualising text. Theme has meaning with respect to a pattern of Themes; New has meaning with respect to a pattern of News; Theme and New have meaning in complementary relation to each other, as part of these complementary patterns (Martin 1992b,c). In some texts, this complementarity may be foregrounded to the extent that attendant ideational and interpersonal meanings are elided (the relevant part of the text is in boldface below):

> For one thousand years, whales have been of commerical interest for meat, oil, meal and whalebone. About 1000 A.D., whaling started with the Basques using sailing vessels and row boats. They concentrated on the slow-moving Right whales. As whaling spread to other countries, whaling shifted to Humpbacks, Grays, Sperms and Bowheads. *By 1500, they were whaling off Greenland; by the 1700s, off Atlantic America; and by the 1800s, in the south Pacific, Antarctic and Bering Sea.* Early in this century, the Norwegians introduced explosive harpoons, fired from guns on catcher boats, and whaling shifted to the larger and faster baleen whales. The introduction of factory ships by Japan and the USSR intensified whaling still further... (W.R. Martin 1989:1)

This foregrounded periodic structure is outlined in Table 2. Note that the Themes participate in a more global pattern of Themes which takes location in time as the principle by which the text orients readers to its field; similarly, the News participate in a more global patterns of News which takes location in space as one principle by which the text elaborates the field as news. These global patterns are further explored in Section 3.3.

Table 2. Ellipsis of other than Theme and New in the whaling recount

By 1500	they	were whaling	off Greenland
by the 1700s	-	-	off Atlantic America
by the 1800s	-	-	in the South Pacific, Antarctic and Bering Sea
Theme [marked]	**Rheme**		
Given			**New**

3. Modes of Meaning and Realisation at the Text Level

In this section we explore the particulate, prosodic and periodic structuring principles in relation to text structure, following up suggestions by Halliday (1981a, 1982) about the ways in which a text is like a clause.

3.1 Prosody

A range of interpersonal meanings at the level of text is explored in (Martin 1992b). The parameter *affect* is taken up here to illustrate prosodic text structure (Martin in press, to appear). In English, *affect* is deployed to negotiate solidarity with the listener/reader. It is an invitation to empathise, which if taken up constructs intimacy and if refused constructs distance. In the following text, a sixteen-year-old secondary school student attempts to share a personal response to a short narrative with her examiner. Interestingly enough, the attempt constitutes a misreading of the examination context by the student, as the examiner's comment reveals. The mark E- constructs maximal social distance for purposes of this public evaluation.

["This response has attempted to give a personal reaction to the question asked. The student has concentrated on the literary style of the story but has failed to answer the question or show any understanding of the story." E-]

The author has intentionally written the ending this way to create the effect that she WANTED [frustration: desire]. I felt EERIE [insecurity: disquiet] and ISOLATED [insecurity: disquiet] after reading the ending—"like a padlock snapping open" sounded so LONELY [insecurity: disquiet] and made me feel so AFRAID [insecurity: apprehension].

I also felt very EMPTY [discord: misery] after reading the passage. It has such a DEPRESSING [discord: misery] ending that made me feel AFRAID [insecurity: apprehension] and SCARED [insecurity: apprehension]. The way "Click" is written by itself in a sentence and in capital letters added to the EMPTINESS [discord:misery] I can really imagine the exact sound it makes, the way it "sounded through the room." "Sounded through the room" is another example of how the author creates the feeling of ISOLATION [insecurity: disquiet] so carefully displayed. It sounds HOLLOW AND DEAD [t-insecurity: disquiet] and creates FEAR [insecurity: apprehension] in your mind.

This is what makes the passage so effective—the way the mood of the characters is portrayed so clearly. I ENJOYED [happiness:care] this passage immensely the ending was very clear and well written.

This text's key is overwhelmingly negative, with only the strangely counterpointed *I enjoyed the passage immensely* to counter the prosody of insecurity and discord. Basically the student's message is an interpersonal one, which is textured in clause final position as news (cf. Section 3.3). And the text is constructed in such a way that the negative response can be made over and over again (disquiet—5 tokens; apprehension—4 tokens; misery—3 tokens), with respect to the story as a whole and to different aspects of its realisation. Through this affectual prosody the student is attempting to construct the examiner as a co-feeler—as someone who is emotionally sensitive to the passage in the same way she is. Unfortunately for the student, emotional solidarity of this kind is not what the examiner is looking for.

3.2 Particle (nuclearity)

At the level of text structure, we explore the orbital/serial interpretation of particulate structure introduced above. This account opposes mono-nuclear texts structures (orbital) to multi-nuclear ones (serial). In terms of representation, orbital structure lends itself to dependency notation (Figure 6), and serial structure to interdependency (Figure 8). Orbital structure is relatively synoptic; it demands that a text hang together around a given centre. Serial structure is relatively dynamic; it allows a text to flow indefinitely from one point to the next. This means that orbital structures are likely to predominate in writing, where sufficient consciousness of text as object can be brought to bear to focus the text, whereas serial structures are likely to predominate in speaking, where it is hard to predict what will happen next and tie everything to a predetermined core. For discussion of synoptic and dynamic structure in relation to literacy and oracy, see (Beaman 1984), (Martin 1985), (Halliday 1985b), (Biber 1988), (Halliday & Martin 1993). This model of particlate structure, based on the notion of nuclearity, is outlined in Figure 11.

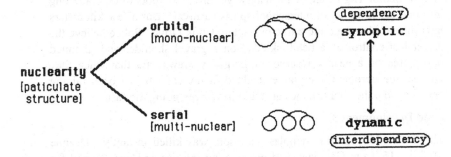

Figure 11. Nuclearity as a model of particulate (ideational) meaning

3.2.1 Nucleus/satellite (experiential metafunction)

The following news story from (Iedema et al. 1994) provides a clear example of orbital structure. The story begins with a Headline, which is elaborated in the Lead. Following this the Lead is elaborated three times, unpacking different aspects of the car crash. In interpreting this text we can treat the Headline and

Lead as nuclear, establishing the focus of the story. Lead Developments 1, 2 and 3 then function as satellites, each one elaborating the nucleus. The crucial point is that Lead Developments 2 and 3 relate to the Lead/Headline in the same way that Lead Development 1 does; Lead Development 2 does not follow on from Lead Development 1, nor does Lead Development 3 from 2. This nucleus/satellite structure is outlined in Figure 12.

Headline

School Jaunt Ends in Death Crash

Lead

A 17-year-old boy was killed instantly when a car carrying eight school friends—two in the boot—skidded on a bend and slammed into a tree yesterday.

Lead Development 1

A 16-year-old girl passenger was in critical condition last night—police said she might have her leg amputated—and a 17-year-old boy was in a serious but stable condition after the tree embedded itself in the car. Incredibly, the two girls in the boot of the V8 Holden Statesman and another girl escaped with only cuts and bruises.

Lead Development 2

The eight friends, two boys and six girls from years 11 and 12, had left Trinity Senior High School in Wagga yesterday at lunchtime, cramming into one car to go to an interschool sports carnival. But a few kilometers later the car ploughed into a tree in Captain Cook Drive. Police believe the driver lost control on a bend, skidded on a gravel shoulder and slammed into a tree on a nearby reserve. Emergency crews said that when they arrived, the uprooted tree was embedded in the car. It had been raining heavily and police believe the car might have been going too fast.

Lead Development 3

The driver, 17-year-old Nicholas Sampson, was killed instantly. Deanne McCaig, 16, from Ganmain, had massive leg injuries and was trapped for more than 90 minutes. She was in a critical condition last night at Wagga Base Hospital, where police say she is in danger of having her leg amputated. Peter Morris, 17, from Coolaman, suffered multiple injuries and was in a serious but stable condition. Among the other students Paulette Scamell and Anita McRae were also in a stable condition, while Shannon Dunn, Catherine Galvin and Rochelle Little, all 16, suffered minor injuries. Police believe the friends from the Catholic high school were on their way to one of the student's homes before heading to the carnival.

[Shelley-Anne Couch, *Sydney Morning Herald* 14/8/92]

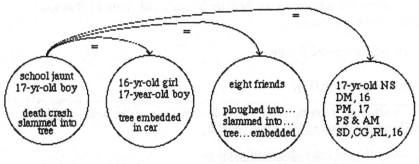

Headline=Lead   Lead Development   Lead Development   Lead Development

Figure 12. Orbital (mono-nuclear) structure in a news story

To confirm this analysis, the way in which information is developed in the news story can be examined in more detail. In the text, information is introduced, to be taken up later with added detail. This uptake and specification is outlined for the eight school friends:

i.   **eight school friends**

ii.   The eight friends, two boys and six girls from years 11 and 12,

i.   A 17-year-old boy was killed instantly

ii. The driver, 17-year-old Nicholas Sampson, was killed instantly.

i.   A 16-year-old girl passenger was in critical condition last night—police said she might have her leg amputated

ii.   Deanne McCaig, 16, from Ganmain, had massive leg injuries and was trapped for more than 90 minutes.

iii. She was in a critical condition last night at Wagga Base Hospital, where police say she is in danger of having her leg amputated.

i.   - and a 17-year-old boy was in a serious but stable condition

ii.   Peter Morris, 17, from Coolaman, suffered multiple injuries and was in a serious but stable condition.

i.   Incredibly, the two girls in the boot of the V8 Holden Statesman and another girl escaped with only cuts and bruises.

ii.   while Shannon Dunn, Catherine Galvin and Rochelle Little, all 16, suffered minor injuries.

i.   Among the other students Paulette Scamell and Anita McRae were also in a stable condition,

The crash itself is handled five times:

ends in death crash (x5):

i.   when a car carrying eight school friends—two in the boot—skidded on a bend and slammed into a tree yesterday.

ii.   after the tree embedded itself in the car.

iii. But a few kilometers later the car ploughed into a tree in Captain Cook Drive.

iv. Police believe the driver lost control on a bend, skidded on a gravel shoulder and slammed into a tree on a nearby reserve.

v.   Emergency crews said that when they arrived, the uprooted tree was embedded in the car. It had been raining heavily and police believe the car might have been going too fast.

And the school jaunt is treated twice:

school jaunt (x2):

i.   had left Trinity Senior High School in Wagga yesterday at lunchtime, cramming into one car to go to an interschool sports carnival.

ii.   Police believe the friends from the Catholic high school were on their way to one of the student's homes before heading to the carnival.

Analysis of this kind reveals the orbital nature of the text structure, with satellites experientially anchored in a Headline/Lead core. The pattern of participant identification (Martin 1992a) in the text confirms the relative independence of one satellite from another. Note that the tree which the car crashes into is introduced in the nucleus as *a tree*, picked up anaphorically in satellite 1 as *the tree*, then reintroduced in satellite 2 as *a tree*, before being picked up in the same satellite as *the uprooted tree*. Satellite 1, which expands the nucleus, treats the tree as recoverable; whereas satellite 2, which does not expand satellite 1, treats the tree as non-recoverable. Note that satellite 2 could have treated the tree as recoverable from the nucleus, but chose not to, apparently preferring to promote its independence of satellite 1 over its connectedness to the nucleus.

Orbital structures are well adapted to newspaper reporting in two main ways. Textually, they make it easier to adapt news stories to the amount of space available; satellites which are relatively independent of each other and which simply elaborate the Lead are easy to prune. Interpersonally, orbital structures make it easier to highlight the potential impact of a story, up front, where it can grab attention; news services owned by just a few magnates depend on highlighting of this kind to attract and control a general readership (for discussion of this literacy evolution, cf. Iedema et al. 1994).

3.2.2 Multi-nuclear (logical metafunction)

Orbital structure's complement, serial structure, is illustrated in the following news bulletin, taken from the television program ZOO TV featuring U2, first broadcast in Australia in 1992. The program is based on U2's Zoo TV tour—a media focussed, multi-textured extravanganza through which U2 moved rock music into a post-modern performance space. The news bulletin was designed to deconstruct representational theories of meaning for a popular audience, and to replace them with theories based on intertextuality and reading position.

> Good morning. I'm Rex Fox for ZOO News in New York. The category is athletics. Born in Czechoslovakia in 1911 from the free games card, the antibiotics arrived too late for 1000's of satisfied motorists. An all night vigil by protestors met with a year's free subscription. Call toll free for ex-government salad sandwiches with a choice of fillings. Older ladies may prefer Beaver playing 'The Decade of Dance' in a crisis currency debate. All these our stories... *ZOO TV News* (ZOO TV featuring U2; 1992)

Visually, the text is constructed as a news bulletin, with an announcer sitting behind a desk in a tv studio. Textually and interpersonally, the language is appropriately contextualised—the bulletin begins with a familiar greeting and identification (*Good morning, I'm Rex Fox...*) and ends with a familiar closing (*All these our stories...*); and it is both authoritative and relatively free of personal evaluation. For most readers, it would be impossible not to recognise the text as instantiating a news bulletin genre. Ideationally, however, the text challenges a mainstream reading position. Although the group and phrase rank meanings are appropriate, the text structure within and between clauses defies expectations. The links between topics are often tenuous, and have to be filled with reference to related news. Figure 13 contains an informal analysis of the text.

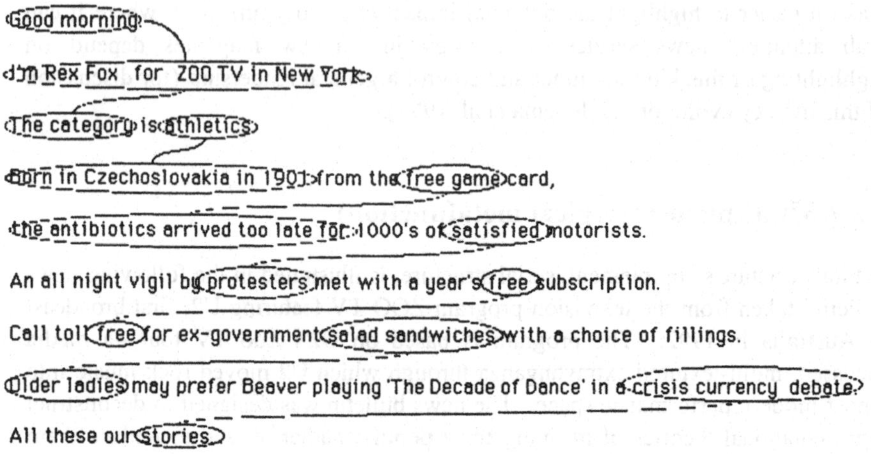

Good morning –

I'm Rex Fox for ZOO TV in New York.

The category is athletics.

Born in Czechoslovakia in 1901, from the free game card,

the antibiotics arrived too late for 1000's of satisfied motorists.

An all night vigil by protesters met with a year's free subscription.

Call toll free for ex-government salad sandwiches with a choice of fillings.

Older ladies may prefer Beaver playing 'The Decade of Dance' in a crisis currency debate.

All these our stories.

Figure 13. Serial progression (via collocation) in the Zoo TV news bulletin

The most revealing way to read a text of this kind is by relating it 'laterally', phrase by phrase to relevant intertexts. The text deliberately does not construct a coherent 'underlying' message of its own; the modernist metaphor of 'depth' thus is erased, as is the duality of meaning and form. In this process, the text deconstructs the notion that meanings lie behind wordings, that news reports on what is going on, that semiosis represents reality and so on. Derrida's (e.g., 1974) transcendental signified is in a sense elided in order to emphasize the fact that all texts, not just post-modern ones, are read against the intertexts that readers from different subject positions bring to bear. In SFL terms, the message is that the ideational metafunction construes reality from the point of view of the speaker/writer in negotiation with the listener/reader. In this process the text itself is never more than a meaning potential, interpellating power (with respect to generation, gender, ethnicity and class).

From the point of view of the general issues under discussion here, the significance of this text lies in its radical foregrounding of serial structure. The text is multi-nuclear; topics are introduced one after another, with no dependent elaboration. Such text structure highlights the dynamic potential of semiotic systems to develop meanings progressively from one point to the next, without any long term regard for what meaning has been made or is yet to come.

3.3 Periodic (textual metafunction)

In light of the distinctive interpretation of periodic structure in SFL, and its relevance to discourse analysis, the personal response text introduced in Section 3.1 is reviewed here to illustrate Halliday's 1985a) analysis of *information* and *theme* in English. The text, represented below, is divided into ranking clauses; ranking clauses are clauses which are not embedded, and which may or may not enter into interdependency relations[4] with adjacent clauses (for this presentation embedded clauses have been included in double square brackets).

ranking clauses:

The author has intentionally written the ending this way
to create the effect [[that she wanted]].
I felt eerie and isolated
after reading the ending—
"like a padlock snapping open" sounded so lonely
and made me feel so afraid.
I also felt very empty
after reading the passage.
It has such a depressing ending
that made me feel afraid and scared.
The way [["Click" is written by itself in a sentence and in...]] added to the emptiness
I can really imagine the <u>exact</u> sound [[it makes]], the way [[it "sounded through the room."]]
"Sounded through..." is another example of [[how the author... isolation [[so ...displayed]]]].
It sounds hollow and dead
and creates fear in your mind. This is [[what makes the passage so effective]]—
the way [[the mood of the characters...]].
I enjoyed this passage immensely
the ending was very clear and well written.

The non-ranking (i.e., embedded clauses) in the text are listed below. For purposes of theme analysis they can be usefully set aside on the grounds that one of the reasons they have been embedded is to displace them from the global thematic organisation of the text (cf. Fries 1981/1983; Martin 1992a, 1993).

embedded clauses:

the effect [[that she wanted]].
The way [["Click" is written by itself in a
sentence and in capital letters]]
the <u>exact</u> sound [[it makes]],
the way [[it "sounded through the room."]]

---

[4] Note that ranking adverbial clauses (e.g., *after reading the ending*), whether finite or non-finite, are often grouped with embedded clauses under the heading subordination in formal grammars (cf. Beaman 1984, Biber 1988, Matthiessen and Thompson 1989 on subordination).

another example of [[how the author creates...
feeling of isolation [[so carefully displayed]]]].
[[what makes the passage so effective]]
the way [[the mood of the characters is portrayed
so clearly]].

Of the ranking clauses in the text, the following nonfinite clauses appear without explicit Subjects:

nonfinite clauses without Subjects:

() to create the effect [[that she wanted]].
after () reading the ending—
after () reading the passage.

And the following paratactically interdependent clauses also appear without explicit Subjects:

branched paratactic clauses (without Subjects):

and made me feel so afraid.
and creates fear in your mind.

For purposes of theme analysis the elided Subjects in these clauses could be filled in; here, however, we will proceed on the assumption that their Subjects have been elided in order to downplay their contribution to the thematic development of the text. This leaves us with the following ranking clauses with explicit Subjects. All these clauses are declarative, in which mood the unmarked Theme is the Subject (following (Halliday 1985a); Themes underlined below).

ranking clauses with explicit Subjects:

The author has intentionally written the ending this way
I felt eerie and isolated
"like a padlock snapping open" sounded so lonely
I also felt very empty
It has such a depressing ending
that made me feel afraid and scared[5].
The way [["Click" is written by itself in a sentence and in...]] added to the emptiness
I can really imagine the exact sound [[it makes]], the way [[it "sounded through the room."]]
"Sounded through..." is another example of [[how the author... isolation [[so ...displayed]]]].
It sounds hollow and dead
This is [[what makes the passage so effective]]—
the way [[the mood of...]]
I enjoyed this passage immensely
the ending was very clear and well written.

---

5 Taken here as an elaborating dependent clause (=b), not an embedded relative.

We are now in a position to characterise what is referred to as the *method of development* of the text (Fries 1981/1983)—the pattern of Themes that construes its perspective on its field. If we restrict our analysis to ranking clauses with explicit Subjects, then the text's angle on its field is two-fold. One aspect is the **student critic**, with subject "I...I...I...I". The other is the text itself, either references to it as a piece of semiosis, or direct quotations from it:

the text:

it (the passage)

that (the passage)

The way "Click" is written by itself in a sentence and in capital letters

this (the way the mood of the characters is portrayed so clearly)

the ending

quotations from text:

"like a padlock snapping open"

"Sounded through the room"

it ("sounded through the room")

These two motifs exhaust thematic selections in the passage, except for *the author,* which might be related to the text reference group.

The motifs associated with New in the text are quite different. For this analysis it is useful to separate ranking from embedded clauses, since ranking clauses are more likely to have their own tone group than embedded ones and so more likely to contribute relatively significant news. Ranking clauses in the text are listed below, with their final clause constituent underline. This analysis assumes a spoken reading of the text in which the tonic syllable falls on the last salient syllable of each ranking clause; this would make at least the final constituent of the clause New (following Halliday 1967, 1970, 1985a,b).

The author has intentionally written the ending this way

I felt eerie and isolated

after reading the ending—

"like a padlock snapping open" sounded so lonely

and made me feel so afraid.

I also felt very empty

after reading the passage.

It has such a depressing ending

that made me feel afraid and scared.

The way [["Click" is written by itself...]] added to the emptiness

It sounds hollow and dead

and creates fear in your mind.

I enjoyed this passage immensely

the ending was very clear and well written.

The overwhelming pattern here has to do with the writer's feelings; it is her personal response to the narrative that counts as news—techincally this is the *point* of her text (Fries 1981/1983). On a slightly more liberal reading of New, this motif could be expanded to include *fear in your mind* and *enjoyed this passage immensely.*

personal reaction:
>eerie and isolated
>so lonely
>so afraid
>very empty
>such a depressing ending
>afraid and scared
>the emptiness
>hollow and dead
>(fear) in your mind

Turning to the text's embedded clauses, where news is arguably less foregrounded, a rather different pattern emerges (asterisked embedded clauses are sentence final, or elaborations of sentence final embeddings, and so would count as news in most oral readings of the text):
>the effect [[that she wanted]]*
>The way [["Click" is written by itself in a sentence
>and in capital letters]]
>the exact sound [[it makes]],*
>= the way [[it "sounded through the room."]]*
>another example of [[how the author... the feeling
>of isolation [[so carefully displayed]]]]. *
>[[what makes the passage so effective]]—*
>= the way [[the mood of the characters is portrayed
>so clearly]].*

Here the author's technical expertise is presented as newsworthy. Standing back a little from these selections, the global picture is one in which i. the student herself is point of departure and her feelings are news; and ii. the text is point of departure and its effectiveness is news—with the first periodic motif predominating. A synopsis of this analysis is presented in Table 3.

4. Conclusion: Types of Structure

In this paper we have followed up suggestions by Halliday about modes of meaning, types of structure, and some ways in which a text is like a clause. The main lesson is that *a text is not a tree*; no form of constituency representation, however elaborate, can respect the complementary particulate, prosodic and periodic structuring principles by which ideational, interpersonal and textual meanings are construed. Secondarily, Martin's 1991 reading of Halliday's 1979

Table 3. Selections for Theme and New in the personal response

Method of development (writer's angle)—unmarked Theme	Point (writer's news)—minimal new
The author	this way
() to	the effect...
[[that	wanted]]
I	eerie and isolated
after ()	the ending
"like a padlock snapping open"	so lonely, so afraid
I	empty
after()	the passage
it (= the passage)	a depressing ending...
[[that	afraid and scared]]
The way "CLICK" is written by itself in a sentence and in capital letters	the emptiness
I	the exact sound...
[[it (= "CLICK")	makes]]
()	the way...
[[it (= "CLICK")	through the room]]
"Sounded through the room"	another example of how...
[[how	the feeling of isolation...
[[()	so carefully displayed]]
it (= "Sounded through the room")	hollow and dead
and ()	fear in your mind
this (= the way the mood ...)	so effective
[[the way	so clearly]]

interpretation of particulate meaning was adjusted, with the segmental part/whole vs part/part reading reworked in terms of nuclearity (orbital vs serial)[6]. This revised association of types of structure with modes of meaning is summarised in Table 4.

The main difference between Halliday's position and that outlined in this paper, is that whereas Halliday appears to associate constituency representation strongly with one mode of meaning (his experiential meaning, part/whole structure association), this paper disassociates constituency representation from any one mode of meaning per se. Rather, as far as ideational meaning is concerned, serial interdependency (logical meaning) is opposed to orbital dependency (experiential

---

[6] I believe that this reading is closer to Halliday 1979: 64–65, who writes "a more appropriate ordering would have a nucleus consisting of a Process plus Goal, with the other elements clustering around it, as in Fig. 6." (a nucleus-satellite diagram).

Table 4. Modes of meaning and types of structure

Types of structure	Mode of meaning
particulate	**ideational meaning**
- orbital [mono-nuclear]	- **experiential**
- serial [multi-nuclear]	- **logical**
prosodic	**interpersonal meaning**
periodic	**textual meaning**

meaning). This structural complementarity can be seen in the nominal group *those two old school friends there from Sydney I told you about that I wanted you to meet.* A structure of this kind foregrounds serial interdependency[7] leftwards from the head, with each segment interpretable as classifying those to its right: *(those (two (old (school friends)))).* To the right of the head however, the structure is orbital, with each segment describing the nucleus *friends: (friends (there) (from Sydney) (I told you about) (that I wanted you to meet)).*

A revision of this kind amounts to a deconstruction of constituency representation as a kind of metafunctional compromise, in which modes of meaning and complementary types of structure tend to be neutralised. What seems to be going on here in English is that a textual wave defines and gives prominence to beginning and end segments; this accounts for the bounded left to right display of constituency representation. Alongside this, experiential nuclearity promotes one segment, the process, as one centre of gravity—typically held responsible for its ensuing complementation. And in addition, in English, interpersonal meaning invests in the Subject and Finite elements, construing an additional centre of attention to the left of the process[8]. These complementary factors give shape and credibility to the constituency tree, however labelled in terms of function and class,

---

[7] Note that giving a multivariate Deictic Numerative Epithet Classifier Thing Qualifier Qualifier Qualifier Qualifier interpretation to the group, alongside a univariate e d g b a b (1 2 3 4) reading (following Halliday 1985a), does not accommodate this serial/orbital opposition.

[8] From this apparently flows the often ethnocentrically farcical search for Subjects in non-Western European languages.

and however branched. English graphology reassures linguists that this form of representation must be foundational and essentially correct, since it has steadily evolved to foreground segmentation over structure of other kinds (Halliday 1985b). While many would acknowledge that this writing system effaces textual and interpersonal meaning, it is not so readily acknowledged that the system of representation propped up by this form of transcription has exactly the same weaknesses. The representational metalanguage (constituency representation) has been shaped by the written language (English graphology) in just this way.[9]

It would seem to follow from these remarks that linguistic theory needs to metastabalise *beyond merocentrism* (i.e., theoretical obsession with segmentation), treating constituency (i.e., one kind of particulate segmentation) not as a primitive, but as a structurally reductive (and experientially biassed) form of representation, the privileged status of which has to do with the evolution of writing systems, not the structure of language. This is not to argue that 21st century grammarians and discourse analysts won't find a place for multivariate representation; there are contexts in which a reductive shorthand can play a productive role (take for example the introduction of generic structure to primary school children; e.g., Christie et al. 1992). But it is to argue that linguists can be more than metascribes, and their theory more than metascription—that a more productive metalanguage can be constructed around the notions of complementary modes of meaning (metafunctions) and of structural configurations (particulate, prosodic and periodic), and that such a linguistics will be exportable across strata, and across semiotic systems, in ways that have not been managed in simple constituency terms.

References

Bateman, J.A. 1989. Dynamic systemic-functional grammar: a new frontier. *Word* 40(1–2) (Systems, Structures and Discourse: selected papers from the Fifteenth International Systemic Congress), 263–286.

Beaman, K. 1984. Coordination and subordination revisited: syntactic complexity in spoken and written narrative discourse. In D. Tannen (ed.), *Coherence in Spoken and Written Discourse*. Norwood, NJ: Ablex. 45–80.

Belsey, C. 1980. *Critical Practice*. London: Methuen.

Biber, D. 1988. *Variation across Speech and Writing*. Cambridge: Cambridge University Press.

Christie, F., B. Gray, P. Gray, M. Macken, J.R. Martin, and J. Rothery. 1992. *Exploring Explanations about Natural Disasters (level 1)*. Sydney: Harcourt Brace Jovanovich (HBJ Language: a resource for meaning).

---

[9] One cannot help wondering if it is not this profound and genrally unacknowledged influence of writing on theory that has overdetermined linguists' pronouncements on the priority of speech over writing, as deconstructed by Derrida (1974).

Cranny-Francis, A. 1990. *Feminist Fiction: Feminist Uses of Generic Fiction.* Cambridge: Polity.

Cranny-Francis, A. 1992. *Engendered Fictions: Analysing Gender in the Production and Reception of Texts.* Sydney: New South Wales University Press.

Derrida, J. 1974. *Of Grammatology.* (translated by G.C. Spivak). Baltimore: John Hopkins University Press.

Fishman, R. 1990. Why Fenech handles himself like a champion in and out of the ring. *Sydney Morning Herald.* Monday May 7, 54.

Fries, P.H. 1981. On the status of theme in English: arguments from discourse. *Forum Linguisticum* 6(1) 1–38. Republished in J.S. Petofi & E. Sozer (eds.), *Micro and Macro Connexity of Texts,* 1983. Hamburg: Helmut Buske Verlag (Papers in Textlinguistics 45), 116–152.

Halliday, M.A.K. 1967. *Intonation and Grammar in British English.* The Hague: Mouton.

Halliday, M.A.K. 1970. *A Course in Spoken English: Intonation.* London: Oxford University Press. Materials accompanied by an audio tape.

Halliday, M.A.K. 1974. Interview with M.A.K. Halliday. In H. Parret (ed.), *Discussing Language.* The Hague: Mouton (Janua Linguarum Series Maior 93), 81–120.

Halliday, M.A.K. 1978. *Language as a Social Semiotic: The Social Interpretation of Language and Meaning.* London: Edward Arnold.

Halliday, M.A.K. 1979a. Modes of meaning and modes of expression: types of grammatical structure, and their determination by different semantic functions. In D.J. Allerton, E. Carney, and D. Holdcroft (eds.), *Function and Context in Linguistic Analysis: essays offered to William Haas.* Cambridge: Cambridge University Press. 57–79.

Halliday, M.A.K. 1979b. Differences between spoken and written language: some implications for literacy teaching. In G. Page, J. Elkin, and B. O'Connor (eds.), *Communication through Reading: proceedings of the Fourth Australian Reading Conference.* Vol. 2. Adelaide, S.A.: Australian Reading Association. 37–52.

Halliday, M.A.K. 1981a. Text semantics and clause grammar: some patterns of realisation. J.E. Copeland, and P.W. Davis (eds.), *The Seventh LACUS Forum.* Columbia, S.C.: Hornbeam Press. 31–59.

Halliday, M.A.K. 1981b. Types of Structure. In M.A.K. Halliday and J.R. Martin (eds.), *Readings in Systemic Linguistics.* London: Batsford. 29–41.

Halliday, M.A.K. 1981c. Structure. In M.A.K. Halliday and J.R. Martin (eds.), *Readings in Systemic Linguistics.* London: Batsford. 122–131.

Halliday, M.A.K. 1982. How is a text like a clause? In *Text Processing: text analysis and generation, text typology and attribution* (Proceedings of Nobel Symposium 51). S. Allen (ed.), Stockholm: Almqvist & Wiksell International. 209–247.

Halliday, M.A.K. 1985a. *Introduction to Functional Grammar.* London: Edward Arnold.

Halliday, M.A.K. 1985b. *Spoken and Written Language.* Geelong, Victoria: Deakin University Press (republished by Oxford University Press 1989).

Halliday, M.A.K. and J.R. Martin. 1993. *Writing Science: Literacy and Discursive Power.* London: Falmer (Critical Perspectives on Literacy and Education) and Pittsburg: University of Pittsburg Press (Pittsburg Series in Composition, Literacy, and Culture).

Iedema, R., S. Feez, and P. White. 1994. *Media Literacy (Literacy in Industry Research Project—Stage 2).* Sydney: Metropolitan East Disadvantaged Schools Program.

Longacre, R.E. 1976. *An Anatomy of Speech Notions.* Lisse: Peter de Ridder.

Longacre, R.E. 1979. The paragraph as a grammatical unit. In T. Givón (ed.), *Syntax & Semantics vol. 12: Discourse and Syntax.* New York: Academic Press.

Martin, J.R. 1985. Process and text: two aspects of semiosis. In J.D. Benson and W.S. Greaves (eds.), *Systemic Perspectives on Discourse vol. 1: Selected Theoretical Papers from the 9th International Systemic Workshop.* Norwood, NJ: Ablex. 248–274.

Martin, J.R. 1990. Interpersonal grammatization: mood and modality in Tagalog. *Philippine Journal of Linguistics* 21(1) (Special Issue on the Silver Anniversary of the Language Study Centre of Philippine Normal College 1964–1989—Part 2). 2-51.

Martin, J.R. 1991. Intrinsic functionality: implications for contextual theory. *Social Semiotics* 1(1), 99–162. Available from Social Semiotics, Department of English, University of Sydney, Sydney NSW 2006, Australia.

Martin, J.R. 1992a. *English Text: System and Structure.* Amsterdam: Benjamins.

Martin, J.R. 1992b. Macroproposals: meaning by degree. In W.C. Mann and S.A. Thompson (eds.), *Discourse Description: Diverse Analyses of a Fund-Raising Text.* Amsterdam: Benjamins. 359–395.

Martin, J.R. 1992c. Theme, method of development and existentiality: the price of reply. *Occasional Papers in Systemic Linguistics* 6, 147–184.

Martin, J.R. 1993. Life as a noun: arresting the universe in science and technology. In M.A.K. Halliday and J.R. Martin (eds.), *Writing Science: literacy and discursive power.* London: Falmer (Critical Perspectives on Literacy and Education). 221–267.

Martin, J.R.1995. Logical meaning, interdependency and the linking particle {-ng/na} in Tagalog. *Functions of Language* 2.2, 189–228.

Martin, J.R. 1996a. Evaluating disruption: symbolising theme in junior secondary narrative. In R. Hasan and G. Williams (eds.), *Literacy in Society.* London: Longman. 124–171.

Martin, J.R. 1996b. Metalinguistic diversity: the case from case. In R. Hasan, D. Butt, and C. Cloran (eds.), *Functional Descriptions: language form and linguistic theory.* Amsterdam: Benjamins (Current Issues in Linguistic Theory). 323–372.

Martin, W.R. 1989. Innovative fisheries management: international whaling. In A.T. Bielak (ed.), *Innovative Fisheries Management Initiatives.* Ottawa: Canadian Wildlife Federation. 1–4.

Matthiessen, C.M.I.M. 1988. Representational issues in systemic functional grammar. In J.D. Benson and W.S. Greaves (eds.), *Systemic Functional Approaches to Discourse.* Norwood, N J: Ablex. 136–175.

Matthiessen, C.M.I.M. 1992. Interpreting the textual metafunction. In M. Davies and L. Ravelli (eds.) *Advances in Systemic Linguistics.* London: Pinter. 37–81.

Matthiessen, C.M.I.M. in press. *Lexicogrammatical Cartography: English systems.* Tokyo: International Language Sciences Publishers.

Matthiessen, C.M.I.M. and J.A. Bateman. 1991. *Text Generation and Systemic Linguistics: experiences from English and Japanese.* London: Pinter.

Matthiessen, C.M.I.M. and J.R. Martin. 1991. A response to Huddleston's review of Halliday's *Introduction to Functional Grammar. Occasional Papers in Systemic Linguistics* 5, 5–74.

Matthiessen, C.M.I.M. and S.A. Thompson. 1989. The structure of discourse and subordination. In J. Haiman and S.A. Thompson (eds.), *Clause Combining in Grammar and Discourse.* Amsterdam: Benjamins. 275–331.

McGregor, W 1990 The metafunctional hypothesis and syntagmatic relations. *Occasional Papers in Systemic Linguistics* vol. 4, 5–50.

Palmer, F.R. 1970. Editor, *Prosodic Analysis.* London: Oxford (Language and Language Learning).

Pike, K.L. 1982. *Linguistic Concepts: an Introduction to Tagmemics.* Lincoln: University of Nebraska Press.

Pike, K.L. and E.G. Pike. 1983. *Text and Tagmeme*. London: Pinter.
Waterson, N. 1956. Some aspects of the phonology of the nominal forms of the Turkish word. *Bulletin of the School of Oriental and African Studies* 18, 578–591 (reprinted in Palmer 1970: 174–187.)

Chapter 3
Interaction and Syntax in the Structure of Conversational Discourse: Collaboration, Overlap, and Syntactic Dissociation

Tsuyoshi Ono[1] and Sandra A. Thompson[2]

[1] University of Arizona

[2] University of California in Santa Barbara

1. Introduction

In this paper we investigate the relationship between interaction and syntax. Using a database of conversational American English, we show how what has traditionally been taken as 'syntax' is intimately involved in the interactional organization of conversational discourse, and we propose a way of thinking about syntax which allows us to integrate the production of syntactic units with interactional structure. We suggest that conversational structure is dependent on a dynamic, interactional notion of syntax. We suggest that examining how syntax works in actual interaction can lead us to a clearer understanding of what syntax is. We hope that an examination of linguistic production in conversation, the most mundane form of linguistic activity, will illuminate the way linguistic resources are exploited in actual production, and that it will further show us how syntactic structures are organized.

There is a growing body of research on the relationship between interaction and syntax. Some of this research has arisen directly within, or has been strongly influenced by, research in the tradition of conversation analysis. Schegloff (1979), in a seminal paper considering the syntactic regularities of repairs, was perhaps the first to propose a 'syntax-for-conversation,' on the grounds that "conversation is the most common and, it would appear, the most fundamental condition of 'language use' or 'discourse'" (p. 283). Goodwin (1979, 1980, 1981) has called attention to the way in which speakers construct, extend, or redesign a sentence in response to interactional demands. Lerner (1987, 1989, 1991, in progress, and to appear) shows how collaborative turn sequences, that is, turn units co-produced by two or more speakers, provide evidence for structures which are projected beyond the first point of syntactic completion, such as contrasts, conditionals, and three-part lists, and Ferrara (1992, 1994) examines collaborative sequences in therapeutic interactions. Fox (1987) shows how interactional factors defining 'adjacency pairs' influence the discourse units that speakers jointly construct within which anaphora operates. Goodwin (1981: chap. 6), Fox and Jasperson (to appear), Fox, Jasperson, and Hayashi (to appear), and Geluykens (to appear) have investigated what repairs and reformulations can tell us about the nature of turn units. Schegloff (1996) provides an appealing view of what a turn-oriented 'sequential syntax' would look like, that is, how turn units might be affected by the interactional organization of conversation.

Other research on interaction and syntax comes from linguistics. Studies of particular grammatical phenomena in their interactional contexts have been the subject of a number of works (e.g., Ashby 1988, 1992, Bentivoglio 1992, Chafe 1987, 1988, 1993, 1994, Duranti 1994, Duranti and Ochs 1979, Ford 1993, Fox 1987, Fox and Thompson 1990a,b, Geluykens 1987, 1989, 1992, Goodwin and Goodwin 1987, 1992, Keenan and Schieffelin 1976, Lambrecht 1984, 1986, 1987, Maynard 1989, Ono and Suzuki 1992, Ono and Thompson 1994, 1995a, Pawley and Syder 1983, Thompson and Mulac 1991a,b, and Weber 1993). There has also been research on implications for discourse structure. In an unpublished but widely circulated work, Pawley and Syder (1977) have provided evidence for a hypothesis that conversationalists operate with a 'one clause at a time' constraint; their data show convincingly that conversationalists find it difficult or impossible to plan the lexical content of a novel sequence longer than a single clause. And Chafe (1987, 1988, 1993, 1994), Du Bois and Schuetze-Coburn (1993), Ford (1993), Iwasaki and Tao (1993), Schuetze-Coburn (1992, to appear), and Tao (1992, 1993) have examined the relationship between clause structure and intonation units in conversation (see below for a definition of intonation unit).

Finally, a collection of work on interaction and grammar from a variety of perspectives can be found in (Ochs et al. to appear).

In this paper, as part of this ongoing program investigating syntax in conversation, we wish to bring together two prominent strands of research into interaction and grammar, one arising from conversation analysis and one arising from discourse/functional linguistics, in addressing the question of the types of interactional phenomena involved in the production of syntactical units.[1] Many of these phenomena have been insightfully discussed in the conversation analytic literature; here we will be considering these questions from the point of view of syntax in an attempt to determine just what kinds of 'knowledge' and 'actions' a model of 'online' syntax might have to account for. In this way, we hope to be led to a closer understanding of the nature of the syntactic resources we bring to conversational interaction, as well as principles underlying the production of syntax and its relationship with interactional factors in actual language use.

---

[1] Several of the examples discussed in this paper are examined in (Ono and Thompson to appear a); however, here we are focusing on different aspects of syntax from those which are the concern of that paper.

2. Analytical Categories

In this section, we present those analytical categories which will be referred to in the body of this paper.

2.1 Intonation Units

Much recent work has shown that spoken language can be fruitfully studied in terms of the basic unit of speech production, a prosodic unit known as the 'intonation unit' (Chafe 1987, 1993, 1994, Du Bois 1991, Du Bois et al. 1993, Du Bois and Schuetze-Coburn 1993, Ford and Thompson to appear, Schuetze-Coburn 1992, and Schuetze-Coburn et al. 1991).
We characterize 'intonation unit' as follows:

> *a stretch of speech uttered under a single coherent intonation contour*
> (Du Bois et al. 1993).

As discussed in (Crystal 1969, Cruttenden 1986, Du Bois et al. 1993, Schuetze-Coburn 1992, to appear, and Schuetze-Coburn et al. 1991), numerous prosodic cues have been identified which can be used to determine intonation unit boundaries. The perception of coherence in the pitch pattern is influenced by both the degree and direction of pitch movement on a stressed syllable and by a change in pitch relative to the speaker's preceding utterance (known as 'pitch reset'). Timing cues also play a role in the perception of intonation units, including an acceleration in tempo on initial unstressed syllables, prosodic lengthening of final syllables, and a noticeable if slight pause between intonation units. The identification of intonation units is thus an auditory, perceptual matter.

Among the basic intonation unit types, there are two that are characterized as ending in a contour which signals finality, designated by a period (final falling) or a question mark (final rising) in the transcription systems of (Chafe 1980, 1987, 1994, Du Bois et al. 1993, and Sacks et al. 1974). In contrast, in all three systems, commas are used to mark non-final intonation contours and dashes are used to mark intonation contours which break off in mid-utterance.[2]

Each of these four types of intonation unit (final falling, final rising, continuing, and truncated) has its characteristic acoustic realizations for a given language (Du Bois et al. 1993).

The intonation unit marked by a period, for English and many other languages, is realized primarily by a marked fall in pitch at the end of the intonation unit. The intonation unit marked by a question mark is realized primarily by a marked high rise in pitch at the end of the intonation unit. There is a wider array of realizations for intonation units marked by the comma, designating a continuing intonation unit

---

[2] As Charles Goodwin has pointed out to us, the truncated intonation unit type plays a very different structural and interactional role from the other three types of intonation units.

type, including a slight rise in pitch at the end of the intonation unit (beginning from a low or mid level), a terminal pitch which remains level, or a terminal pitch which falls slightly but not low enough to be considered final. In practice the comma represents a broad variety of non-final non-truncated contours, whose members may be distinguished to some extent by their terminal pitch direction. As discussed in (Chafe 1994, Du Bois and Schuetze-Coburn 1993, Ford and Thompson to appear, and Ono and Thompson 1995a), and as we see below, 'intonation unit' closely interacts with the realization of syntactic completion to produce the turn units basic to the organization of the structure of conversation.

2.2 Constructional Schemas

Most linguists have taken the 'clause' to be basic in the use of language. Croft (1991:2), for example, suggests that 'the structure of clauses' is central to an understanding of the grammar of human languages. Chafe (1994:66) suggests that the prominence of the notion of 'clause' derives from the fact that events and states are most likely to be verbalized as clauses. Goodwin (1979, 1981), Lerner (1987, 1989, 1991), and Schegloff (1979, 1988, 1991) assume the notion of 'sentence' to be uncontroversial, using it to include what we are calling 'clauses' and 'clause combinations.'

We consider 'clause' and 'clause combination' to be abstract syntactic categories which can be thought of in terms of conventionalized 'constructional schemas' in the sense of (Langacker 1987:2.1,11). That is, constructional schemas are patterns, distilled from large numbers of speech events, to the point where they have a cognitive status independent of any particular context. The grammar of a language can be understood as a structured inventory of such patterns (Langacker 1987:63, Fillmore 1988:37).

Unlike prototypical lexical items, grammatical patterns are schematic rather than specific: that is, they do not represent particular expressions, but are schematizations over sets of expressions parallel in formation, which are their instantiations.   Constructional schemas serve as templates for analogous expressions (Langacker 1987:68, 1991:46). Based on their frequency, patterns are more or less well entrenched (Du Bois 1985, Langacker 1987:59). The more a given pattern is used, the more strongly entrenched it becomes, and the more it becomes 'grammaticized,' a part of the 'grammar' of the language.

Schemas are not only complex, but they vary in size from multiple clauses to phonological combinations; they can be nested and combined in various ways, and speakers' knowledge of these interactions is, of course, also part of their syntactic knowledge. Thus, for example, a clause such as that in (1) below must involve both an 'SV' schema and a 'Prepositional Phrase' schema:

(1) Africa 2

A: ...actually,

they just went out to <% Chisera= %>,[3]

Examples of constructional schemas in English would be patterns for producing agentive nominalizations, 'SV(O)' declarative clauses, passives, WH-questions, and conditional combinations. Schemas can be combined, connected, and embedded in ways that we are not ready to describe, but will hint at in the examples below. As Langacker (1987, 1991) and Fox (1994) have shown, the schemas themselves are far from simple, and cannot be easily described. The category 'relative clause' in English, for example, like most other categories humans attend to, is a complex, perhaps 'radial,' or 'family resemblance,' category (Wittgenstein 1958, Lakoff 1987), with non-discrete boundaries, rather than one whose members can be described in terms of a set of shared properties. For further discussion of schemas in syntax, cf. (Barlow and Kemmer to appear, Langacker 1987, 1991, and Ono and Thompson 1995a).

Having outlined previous research and our assumptions, we proceed to propose an account of the ways in which these constructional schemas and interaction work together to produce 'syntactic structures' in conversation.

3. Discussion

Our discussion begins with the general fact, discussed extensively in the CA literature, that conversation itself is an interactional achievement.[4]   A reasonable corollary of this fact would be that the production of syntactic units is itself also an interactional achievement. Indeed, a number of recent studies have made this very point (see especially Clark and Wilkes-Gibbes 1986, Ferrara 1992, 1994, Ford 1993, Fox and Thompson 1990b, Geluykens 1987, 1989, 1992, Goodwin 1979, 1980, 1981, Goodwin and Goodwin 1987, 1992, Lerner 1987, 1989, 1991, in progress, to appear, and Schegloff 1979, to appear).

Taking these studies as our starting point, we will explore in this section the implications of the profound collaborativeness of syntactic production for our understanding of what a model of syntax should look like. We would like to consider the present effort as only a first step toward more systematic examination of the interface between syntax and interaction.

This discussion is organized so that we consider first those products of collaboration which result in what grammarians would consider to be canonical

---

[3] In our transcription system, adapted from Du Bois (1991) and Du Bois et al. (1993), each line represents one intonation unit. Please refer to the appendix for a summary of the Du Bois et al. transcription conventions.

[4] Nearly every contribution to conversation analysis in the last 20 years has made this point; for this term, see especially (Schegloff 1982, 1988, and 1995), and for explicit discussion of this point, see (Goodwin 1981, Schegloff this volume and to appear).

instantiations of constructional schemas, and then those which result in instantiations which do not match any schema, but which can be understood as instances of syntax being subordinated to interactional demands.

3.1 Collaborative Production of 'Clean' Syntax

One of the most interesting and pervasive phenomena in conversational data is a range of collaborative activities involved in the production of syntactic units; examining our data carefully, we can easily find several types of collaboratively achieved realization of syntax, each of which provides a slightly different perspective on the issue of the co-construction of syntactic units. In this section, we will investigate each of them in turn. These different types of syntactic collaboration reveal a range of syntactic skills which speakers bring to interaction; we suggest that a close investigation of these skills allows a deeper understanding of the nature of syntax itself.

3.1.1 Speaker A Does not Complete a Syntactic Unit

Speaker B directly offers a completion to what Speaker A has started
A frequent and dramatic type of co-construction is that in which Speaker A leaves a syntactic unit unfinished and speaker B finishes it. This type of co-construction has received the most attention in the grammar-and-interaction literature (Ferrara 1992, 1994, Lerner 1987, 1989, 1991, in progress, Ono and Thompson to appear a, and Ono and Yoshida to appear). Here is an example:

> (2) Cuz 31
> L: his position is pretty uh,
> A: ... % (TSK) (H) stable.
>   .. yeah.

In this example, L breaks off before finishing a realization of a predicate adjective clause schema; syntactically, a predicate adjective is called for, but lexically, A has a range of choices which could be appropriate. In interpreting this example, it is useful to know that A has been the primary speaker, talking about 'his position' at work. In this excerpt, then, an opportunity arises (see Lerner to appear) for A to give the syntactically required predicate adjective, and interestingly, A immediately takes the opportunity, not only providing the adjective, but confirming the inference underlying L's unfinished clause, namely her inference from the preceding discussion that 'his position is pretty [positive adjective].'

The following examples further illustrate instances of 'Speaker B finishing what Speaker A has left incomplete':

   (3) Lam 18

	M:	...() (H) are they,
		...() teaching,
		.. any more lambada,
		...() at,
		.. uh,
→	J:	...() school?
	M:	.. yeah.
	H:	...() Vivian was giving a lambada lesson,

In the first five lines of (3), M produces a nearly complete clause, in the form of a question. Syntactically, he has produced everything but the object of the preposition *at*. As can be seen at the arrow, J provides a candidate prepositional object *school* for M's clause, but she does it with an 'appeal' intonation contour (Du Bois' term for question/final-rising contour), which serves interactionally to ask M if this is what he meant. In the following line M indicates his acceptance of the candidate completion.

The next example of co-construction raises an intriguing interactional point:

   (4) Lam 4
(four friends are talking about children's skeletal flexibility; M is an MD)

1 M:	so that's why they always have ... more flexibility.	
2		cause [1 as 1] they're growing,
3 P:		[1 hm 1].
4 M:	...() you're making cartilage.	
5		so it [2 al- —
6 P:		[2 mhm 2].
7 M:		they 2] always have that,
8		until they reach adulthoo=d,
9		... in which case
→ 10 H:	they get old and cranky [3 like the rest of 3] [4 us 4].	
11 M:		[3 it stops 3] [4 = 4].

In (4), M is explaining why children's bones are more flexible than adults'. At lines 7-8 he produces a complex syntactic clause combination, and then in line 9, he begins a non-restrictive relative clause. At the arrow in line 10, H jumps in to provide a joking collaborative completion to the non-restrictive relative clause. In line 11, M provides his own completion to this non-restrictive relative clause. From the tape it is clear that H intends his contribution as a joking *superimposition* to the serious completion which he has every reason to think M will actually provide. In this way, H and M have collaborated to produce two distinct but compatible and to-be-heard completions to the same syntactic unit that M started in line 9.

The following example is intriguing because the completion offered by Speaker B is *not* accepted by Speaker A, though it serves another interactional goal:

(5) Lam 22

J:	(H) maybe Harold you should come with me.
H:	...() why.
	... so I can find —
→ J:	.. learn some [lam-] .. lambada.
P:	[@@]
H:	.. find some girls with empty chairs next to em?

In line 2 of (5), H asks J (who is his wife) why he should come with her to the dancing class. At this point, his question can be seen as either genuine or rhetorical. In the next line, H displays a rhetorical interpretation of his own question, by starting a facetious answer to it, which he leaves incomplete. At the arrow, J, choosing to interpret the question as a genuine request for information, offers a 'completion' for H's incomplete instantiation, but this completion crucially involves 'substituting' the verb *learn* for the verb *find* and thus provides a totally different 'completion' than H himself intended, as becomes clear in the final line of the excerpt, where he finishes what he had presumably started to say. This interaction can be seen at two levels, then. Syntactically, J has provided a 'completion' for H's incomplete clause, and at the same time she has interactionally answered his why-question. H's own next contribution provides an alternative syntactic completion and interactionally allows him to give his facetious answer to his own question. Technically, they have 'collaborated' on producing a syntactic unit, but it is not the syntactic unit that H had actually started.

The next example is similar to (5) in that both participants appear to be providing answers to the same question, but syntactically it raises a difficult issue:

(6) Farm 7
(B has been talking about problems a friend is having with his tractor)

1 L:	...() so what's he thi=nk.
2 B:	...() we=ll,
3 L:	.. going to have to —
→ 4 B:	... he thought,
→ 5	...() re=bui=ld,
6	...and he thought maybe the motor was just wearing out,
7	...it's got so many hours on it,

In the first line of the this excerpt, L poses a question. B begins an answer with *well*, and then in line 3 L starts to provide what might appear to be an answer to his own question. From the preceding context, however, where B has been doing all the talking about the tractor problem, it seems clear that L's *going to have to* is actually better viewed as a 'prompt' for B to continue. What B does next is very interesting: at the first arrow, in line 4, he starts to answer L's original question *so what's he think?* with *he thought,*. In line 5, at the second arrow, however, he breaks off before finishing what it might have been that 'he thought,' interrupts this schema instantiation, and offers what can only be understood syntactically as a completion of L's prompt in line 3:

(7) going to have to—rebuild,

thus accepting that earlier remark as a prompt, and collaboratively finishing it. Like the previous example, this example shows the skill with which speakers are able to manage several interactional goals at once; one outcome is the collaborative production of syntactic units.

What we have considered so far, then, are examples illustrating a type of collaboration in which Speaker B directly finishes a syntactic unit that Speaker A has left unfinished. Next we examine instances in which Speaker B can be said to have been offered an opportunity to help to achieve a syntactic completion.

Speaker A allows an opportunity which Speaker B can take advantage of to complete what Speaker A has left incomplete

We often see collaborations in which Speaker A provides an opportunity for Speaker B to finish what A has started. Sometimes Speaker B takes advantage of the opportunity and sometime s/he doesn't (for further discussion, see (Lerner 1991, to appear)). We'll consider examples of each type.

Speaker B takes advantage of the opportunity

In the following example, Speaker A not only fails to complete a syntactic unit, but interactionally solicits Speaker B's involvement in trying to finish what s/he wishes to express. In our data this is often accomplished by *you know* (Ono and Thompson 1995a):[5]

```
(8) Carsales 12
      D:      ...I just wanted to make sure, [that I] go back,
      G:                                     [you —you-] —
 ⊕  D:      ...and...and and people aren't going to —
              ...you know,
      G:      ...right,
              ...get on your case,
```

At the arrow, D begins to instantiate a schema which he doesn't finish, a schema which can be continued only by a verbal expression:

(9) NP be going to [verbal expression][6]

From the tape we can't tell why D doesn't finish a realization of this schema, but it is clear that his *you know* is an interactional move which allows him several

---

[5] For further discussion of *you know*, see (Östman 1981, Schiffrin 1987).

[6] D's actual utterance also contains a marker of negation. We are considering it to be an instantiation of a separate schema.

options for achieving closure on the realization of this schema: it can provide an opportunity for G to finish what he, D, has started,[7] or as a way of buying time so that D himself can find an appropriate verbal expression. In the event, G takes the opportunity to offer a verbal expression. The realization of this schema, jointly produced by D and G, is highly determined by the interaction of D's inviting G to help him to produce an instantiation of his schema.

Speaker B doesn't take advantage of the opportunity
When Speaker B doesn't take advantage of the opportunity, we have a situation where Speaker A finishes his/her own unfinished syntactic unit after giving Speaker B an opportunity to participate in finishing it. For example:

> (10) Lam 5
> (joking about whether J's friends are going to 'stick up for' her)
>   1 P: ...() I stuck up for you today at that store.
>   2 H: that's true.
>   3 J: ... you did.
> → 4   [you made me get] the,
>   5 P: [mhm],
>   6 J: um,
>   7 P: that's right.
> → 8 J: the green <X scarf X>.
>   9   ...() that's right.

Here in line 4, J starts a syntactic unit but stops after the definite article. From the context it is clear that P, H, and J all know in what way P 'stuck up' for J earlier, so they can all provide the completion for J's clause started in line 4. J hesitates in line 6, but P does not take the opportunity to provide a completion for the NP. Instead, interestingly enough, P acknowledges the correctness of the clause which no one has yet finished with his *that's right* in line 7. Finally, J herself finishes the clause in line 8.

The next example is also of this type: Speaker B refrains from completing Speaker A's unfinished syntactic unit, but instead acknowledges the correctness of the as-yet-unfinished clause:

> (11) Carsales 4
>   1 D: .. she spent twelve years of her life with me,
>   2   and u=h,
>   3   ...() she's always been positive,
>   4   .. thinker,
>   5   and uh,
>   6   .. always been good,
>   7 G: .. yeah,

---

[7] See (Fox and Jasperson to appear) for discussion of similar situations involving invited utterance completions.

8 D:	... understanding,
9	... [and u=h],
10 G:	[(H)] sure,
11	it would be different,
12	if she were a bitch,
13	and always [2 nagging,
14	.. you know,
15	and then 2] .. getting on your case,
16 D:	[2 yea=h,
17	..exactly 2].
18	[3 or didn't like —
19	.. or didn't enjo=y,
20	doing anything 3].
21 G:	[3 and making your life impossible 3].
22 D:	.. she [4 always was,
23	.. you know 4].
24 G:	[4 yeah.
25	..exactly 4].
26 D:	...() (H) pretty much u=h,
27	... able to do anything that I wanted to do.

In this excerpt, D has been talking about some of the ways in which his former girlfriend was good to him, and G has been supportive in building this positive picture of her. In line 22, D begins a syntactic unit which needs a verbal element to complete it: *she always was* ___ . In the next line, D's *you know* can be seen as either creating an opportunity for G to furnish an appropriate completion or allowing more time for him (D) to find one. Again, G does something which is quite remarkable from the point of view of propositional semantics: he offers the reactive tokens *yeah* and *exactly*,*even though* D has not yet said any predicate to which these responses would, strictly speaking, be appropriate. Interactionally, of course, there is nothing strange about this, in that G's reactive tokens can be taken as tokens of appreciation for the general direction in which D's talk has been going. Finally, in lines 26-27, D finishes the syntactic unit which he started in line 22. We interpret this example as a clear illustration of the interplay between interactional and syntactic goals. Interactionally, D and G collaborate to produce a complete syntactic unit in that G's reactive tokens can be seen as a warrant for D to take the time he needs to find an appropriate completion.

In looking at ways in which Speaker A allows an opportunity for Speaker B to complete an incomplete syntactic unit, we have seen that complete units can be collaboratively produced in one of two ways. If Speaker B takes advantage of the opportunity, we have an instance of Speaker A starting a unit, allowing an opportunity for Speaker B to finish it, and Speaker B finishing it. If Speaker B doesn't take advantage of the opportunity, then Speaker A can finish the unit him/herself, with a collaborative warrant from Speaker B.

3.1.2 Speaker A's Unit Becomes Part of a New Unit

Speakers can build new units together by expanding what one of them has said into a new syntactic unit. This can happen in three ways: Speaker B can directly expand Speaker A's unit, Speaker A can expand her/his own unit after input from other speakers, or Speaker B can expand Speaker A's unit after input from other speakers.[8] We'll consider an example or two of each type.

Speaker B directly expands Speaker A's unit
Consider example (12):

(12) Lam 2
    M: ... he was actually here two weeks ago,
          and [I missed him].
    J: [at the .. at] the ja=zz .. tap thing or whatever.

In this example, M produces a complete clause in the first line. He gives evidence that he has more to say in the continuing intonation contour of this clause, indicated by the comma at the end of this line, and indeed, he proceeds to add another clause in the second line. But J overlaps this clause, adding a prepositional phrase to M's clause, which has the effect of turning M's contribution in line 1 into the beginning for a new clause, which is readily interpretable, and is syntactically and intonationally complete at *at the jazz tap thing or whatever*.

Next consider example (13):

(13) Hypo 4
    1 G: .. a=nd,
    2   it ca=n cause ca=ncer.
→ 3 K: ... which he ha=d.
→ 4   .. which he got from it.
    5   .. @@
    6   but he's @since recovered.
    7   ... @N
    8 G: well,
    9   the worst [thing I ever had,
    10   was brai=n] fever,
    11 K: [@N @he's a medical miracle].
    12 G: when I <X had X> [2 proposed   2] to her.
    13 D: [2<P @@@ P>2]
    14 K: .. @@@@
→ 15   ... (H) from which you haven't recovered.

---

[8] Ono and Thompson (1995b) discuss this phenomenon in detail in terms of Langacker's (1987, 1991) notion of 'conceptual structure'.

At the first two arrows, K adds two non-restrictive relative clauses to the clause which G had finished in line 2. The non-restrictive relative clause at line 15 is even more striking, since it goes with the NP *brain fever* in line 10, but occurs after G's own *when*-clause in line 12. In both instances, however, G has produced a complete clause which is turned into the first part of a new clause combination.

As another example, we note a variation on this theme: before adding to Speaker A's complete clause, Speaker B can first acknowledge it:

(14) Hypo 7
      1 D: this explains u=h .. Judaism.
      2 K: ... [not] really.
      3 D: [in] ... big print,
      4   in= —
      5   ... in two hundred and eighteen pages.
      6 K: ... Greg got it.
→ 7 D: ... that's nice.
→ 8   ... for your daughter.

In (14), K produces a complete clause at line 6, a realization of schemas something like:

(15) NP V NP

Before adding a prepositional phrase to that clause, D first acknowledges K's contribution. D's addition in this instance is interesting deictically, in that while the result is a realization of an expanded schema:

(16) NP V NP PP

the deixis in line 8, *your daughter*, is that which is appropriate for the pragmatic relationship between the two speakers, not that which would be appropriate if the entire expanded clause had been uttered by either one of them.

Speaker A expands her/his own unit after input from other speakers

The second type of expansion of a complete syntactic unit is that which is expanded by the same speaker after input from other speakers. In these examples, the collaboration is of a more subtle kind: a new syntactic unit emerges from the *interaction* among the speakers, although the unit is produced by Speaker A. Consider (17):

(17) Farm 2
      L: ...I think it's fifty dollars a day.
      B: ... oh!
      L: ...() to rent one.

In (17), L's input in the first line is a complete syntactic unit. B's *oh!* in the second line provides input to L that an increment (Sacks et al. 1974, Ford 1993, Ford and Thompson to appear, Wilson and Zimmerman 1986) may be warranted, and L adds the increment in line 3 to his already complete syntactic unit.

The following example illustrates a similar process:

(18) Lam 25

   1 J: remember a few months ago I used to go out
              dancing?

   2   .. every now and then?

   3 H: hm-m,

   4   I don't remember.

   5 M: ...() well the thing that gets me,

   6   ...() I meet [this=] —

   7 J: [to Caesar's],

   8   and stuff?

In lines 1 and 2, J reminds her husband H of her earlier dancing activities, which H claims not to remember, as seen in line 4. In line 5, M begins to tell what will turn into an extended story, but J, still trying to remind H about her earlier dancing, specifies one of the actual places where she went out dancing by adding the prepositional phrase *to Caesar's* to her expanded clause in lines 1 and 2. It seems reasonable to suggest that this specification is motivated to remedy J's husband's forgetfulness. Thus we have another example where the actual instantiation of a syntactic unit is shaped by the ongoing interaction.

Here is another example:

(19) Hypo 2

   1 G: ...() I'd like to have .. my% .. lu=ngs,

   2   ... my entire respiratory tract,

   3   ... (H) replaced,

   4   ... (H) with .. asbestos.

   5   .. or something.

   6   .. (H)

   7 K: @N@N@N

→ 8 G: <F or canvas F>.

In lines 1-5 of (19), G produces a complex realization of a set of schemas which we might represent something like this:

(20)   like to Verbal Expression
            to have NP V-ed
            replace NP with NP

with the instantiation of the first NP in the schema being reformulated from *my lungs* to the more specific *my entire respiratory tract* in a separate intonation unit before G goes on to instantiate the rest of the schema. Then, as seen in line 8, G constructs the second NP in the third schema above as an instance of an 'alternative NP' schema of the form:

(21) NP or NP[9]

We suggest that this expansion of the set of schemas in (20) to include that in (21) happens interactively: it is manifestly the case that G adds the realization of the 'alternative NP' schema, *or canvas*, only after K has contributed her appreciation token (Lerner to appear). (19) thus provides another example of interactional factors shaping the way in which syntactic resources are utilized.

These examples show how speakers are able to take a complete syntactic unit of their own as the beginning of a new expanded syntactic unit, made relevant by input from other speakers.

Speaker B expands Speaker A's unit after input from other speakers

We have also found cases where Speaker B expands Speaker A's complete unit after input from other speakers. Consider (22):

> (22) Lam 23
>> (talking about heavy flirting at a dancing place)
>> 1 M: .. I figured,
>> 2 J: [oo=].
>> 3 M: [oh,
>> 4   they must know each] other.
>> 5 J: ...() oo=.
> → 6 H: .. very well.
>> 7   <@ in fact @>.

In lines 1 and 3-4, M produces a syntactically complete utterance, *I figured, oh, they must know each other.* After J's appreciation token *oo=*, H then adds to M's completed syntactic unit with his *very well, in fact.*

A more dynamic instance of this process can be observed in the following example:

> (23) Lam 7
>> 1 H: .. usually we just have r=eally loud salsa parties
>>   across the street.
>> 2 J: that's fun too.
>> 3   ... and teenagers,
>> 4   ... kissing each other on the side[walk].
>> 5 H: [hm=].
>> 6 P: hm=.
>> 7 H: (KISS) (LATERAL_CLICK) (LATERAL_CLICK)
>> 8 P: ...() and little kids throwing p=aint in your
>>   backyar=d.

---

[9] Acknowledging the possibility that it developed from use as an instantiation of the alternation schema, we consider *or something* in line 5 of this excerpt to be a discourse marker and not part of the alternation schema.

    9 H: yeah=.
    10 J: ...() (GROWL)
    11 P: @
    12 J: ...() [(SIGH)]
    13 H: [and in] the front yard-.

H and J are married and are talking about the living environment around their house to their friends P and M (who does not talk in this segment). In line 1, H produces a syntactically complete utterance describing the 'noise' situation around the house. In lines 2-4, J acknowledges it in a sarcastic manner with *that's fun too.* Then she expands H's utterance by saying *and teenagers kissing each other on the sidewalk.* H and J start animatedly describing what is going on around their house and how they really feel about it by making kissing sounds, growling, and sighing as seen in lines 7, 10, and 12. Notice that in line 8, P, who has stayed at H and J's place and experienced the situation, expands J's utterance by saying *and little kids throwing paint in your backyard.* H then expands this expanded utterance by saying *and in the front yard.* H's choice of expression *and in the front yard* makes it clear that he is expanding P's utterance in line 8, to jointly produce yet another new expanded collaborative syntactic unit.

3.1.3 Summary

We have considered several types of examples of syntactic 'co-constructions,' broadly conceived, and suggested that each of these types provides important clues about the interface between syntax and interaction. The production of syntactic units is often a joint activity, strongly suggesting that speakers share not only a knowledge of possible syntactic unit schemas but also a knowledge of how to expand shorter schemas into longer ones. Syntactically, expansions can occur as easily from one's own productions as from another's, and in fact, there are many opportunities in conversational interaction for collaboration in the production of syntactic units.

So far, the cases we have examined have involved ways in which interaction influences syntax with relatively orderly results. Indeed, quite often the results of joint syntactic productions are syntactically parallel to realizations produced by a single speaker. That is, we have examined instances in which speakers collaborate to produce syntactic units which are clear instantiations of well-established schemas. In the next section, we consider instances in which the very interactional nature of conversation produces circumstances in which syntactic units are produced that do not match any schemas.

3.2 'Messed up' Syntax

In examining instances of 'messed up' syntax in conversation, it is important to note that in none of the cases to be reported below do we have evidence that 'messed up' syntax causes any problems for the participants. Needless to say, there are occasions in conversation where it becomes a problem, but the examples below are indicative of the frequency with which syntactic 'problems' for analysts are unremarkable to participants in ordinary conversational interaction.

3.2.1 Overlap

One frequent situation where one finds 'messed up' syntax is where speakers overlap. The sequential implications of overlap resolution have been insightfully discussed in (Jefferson 1973, Schegloff 1987); here we seek to determine what implications overlap might have for syntax.

For example, in the following excerpt, at line 5, D's syntactic unit is broken off after *don't*, presumably because lines 3-5 are in overlap with what G is trying to say in lines 6–9. Only when D stops what he is saying in line 5 is G's syntactic unit actually brought to closure in the clear in lines 10–11, after which D's broken-off clausal instantiation in line 5 is also completed in the clear in line 13:

```
(24)  Carsales 8
     1 D:    y —
     2       y —
     3       [y —
     4        .. you don't think about other things,
     5          .. and you don't] —
     6 G:    [yeah,
     7        it's —
     8         it's —
     9          it%] —
    10       .. it —
    11       it's not a job that requires a lot of thinking.
    12 D:   .. right.
    13      .. and you don't relate to any people.
```

The next two examples involve cases in which two speakers are each trying to add to a previously completed syntactic unit:

```
(25)  Carsales 2
     1 G:    that's cheap man,
     2        because,
     3 D:    ... I just got [liability].
     4 G:                    [Trip-] —
     5        .. Triple A wanted to give me,
```

6   ... you know,
7   .. insurance,
8   .. e%- eleven hundred too man.
9 D: (0) yeah.
→ 10   [for a reg —
11   .. for a Dodge].
12 G: [o% —
13   on a fucking —
14   .. on —
15   .. on%] —
16   .. on an Aries.

In lines 4–8, G describes the high cost of the insurance he was considering. His utterance ends with the 'unattached NP' *eleven hundred too man*-3   (on 'unattached NPs, see Ono and Thompson 1994 and references cited there), which has a final intonation contour.   At this point, both G and D try to add a prepositional phrase to G's completed clause to make a realization of a new expanded clause schema. D's lines 10–11 overlap with G's 12–15; G's response to the overlap is to continue to recycle his turn until D has completed his turn in line 11, and then, with precision timing (Jefferson 1973, Schegloff 1987), he produces the end of his turn in the clear in line 16. Note that both D's contribution in lines 9–11 and G's in lines 12–16 are intonationally and syntactically complete, with 'syntactically complete' crucially being understood in terms of its SEQUENTIAL relationship with the previous (already complete) material.

To describe this situation, paraphrasing Jefferson and Schegloff 1975, we might say that a speaker may manipulate portions of the realization of the syntactic unit s/he is currently producing, allowing them to occur in the clear.

A more complex type of situation is encountered when a syntactic unit is abandoned because of what is heard during overlap. Consider example (26), an extension of (14) considered above:

(26) Hypo 7
       1 D: this explains u=h .. Judaism.
       2 K: ... [not] really.
       3 D: [in] ... big print,
       4   in= —
       5   ... in two hundred and eighteen pages.
       6 K: ... Greg got it.
       7 D: ... that's nice.
    → 8   ... [for your daughter].
    → 9 K: [from the library],
       10   for% —
       11   ... unhunh,
       12 G: .. ((SIPPING_TEA)) .. I want her to know,
       13   what she's rejecting.

In this example, in line 6, K produces a complete clause. Very much as in example (25) just above, in lines 8 and 9, D and K both produce prepositional phrase realizations of expansions to this clause schema in overlap, beginning and ending at almost exactly the same split second. Interestingly, D's contribution constitutes the realization of a new complete clause and is marked with a final intonation contour. But K's addition, while syntactically complete, is marked with a continuing intonation pattern (indicated by the comma), making it clear that she is projecting more talk. And indeed, the effect of the overlap can clearly be seen in lines 10 and 11, where K realizes that D's prepositional phrase was what she was about to say; she starts to say who the book was for in line 10, and then abandons that prepositional phrase with a glottal stop ('%'), and switches to a reactive token *unhunh* that confirms what D had said. So we analysts find a syntactically abandoned utterance. However, of course, the speakers are not bothered by that at all. For them, what emerges interactionally seems to be good enough for them to go on with what they are currently engaged in. That is, the combination of semantic, cognitive, and pragmatic factors wins out over the mere production of syntactically impeccable schema instantiations (Ono and Thompson 1995a).

These examples further illustrate ways in which interaction influences the production of syntax. When speakers overlap, they are aware that their production of syntactic units is being 'interfered with.' The ways in which speakers creatively utilize their syntactic resources to resolve the interactional and syntactic complications which arise as a result of overlap can shed light on the nature of these resources, and the role they play in achieving satisfactory interactions.

3.2.2 Dissociation from Schema Instantiation

In this section we consider the 'messed up' syntax that emerges when a schema instantiation becomes dissociated from the schema of which it was an instance. Let us begin with (27):

```
(27) Hypo 3
     1 K:     .. he's only had,
     2        (H) .. <@ since .. since we've been @> married,
     3        ... ca=ncer,
     4 D:     @@[@@]
     5 K:        [@@]@@
     6        .. (H) .. @leukemia=,
     7        ... (H) bronchitis=,
     8        ... (H) uh=,
     9        .. tuberculo=sis,
    10        .. @@@@ (H)
    11        .. and he's recovered from all of them.
    12        that's what's [so ama=zing].
  → 13 G:                  [nasal polyps],
```

```
      14 K:     @@@ (H)
      15        @and <X @then X>,
      16        (H H)
      17        .. u=h —
   → 18 G:      (0) s=creaming itching assho=le,
      19 K:     .. @@ [(H)]
      20 G:            [that's one] of them.
   → 21 K:      (0) lumba=go=,
```

Semi-jokingly, G has been saying that he has had a range of diseases and health problems. In lines 1 to 9, his wife K lists all the serious diseases that he has supposedly had. She finishes her list in line 9, and adds a syntactically and intonationally complete realization of a co-ordinate clause in line 11. In line 12, she adds a realization of another clausal schema. At this point, G does something which is rather remarkable syntactically but quite unremarkable interactionally: in line 13, overlapping with K's clause, he begins to *add to* the list of diseases which K had presumably finished.

To be interpretable, it seems reasonable to say that G's NP *nasal polyps* is syntactically being taken as an extension of the list of direct objects of *had* in line 1, even though the clause of which that verb is the predicate has been explicitly completed in line 9, the clause combination of which it is the first part has been explicitly completed in line 11, and a syntactically entirely unrelated assessment clause realization has been produced. Even more interesting from a syntactic point of view, K immediately accepts the re-opening of the list of NPs which count as direct objects of *had*, fishing for another candidate in lines 15–17, G then adds a candidate in line 18, and K adds one in line 21. In other words, we can easily say that while the NP schemas G and K are instantiating are interspersed with full clauses (lines 11, 12, and 20), both speakers are jointly producing a coherent instantiation of the 'list schema':[10]

(28) if an NP is appropriate, then a list of NP's is appropriate

It seems that our knowledge of this syntactic generalization extends to applying it even after several signals of clausal closure have occurred.

But at some point around the middle of this excerpt, it also seems reasonable to say that the list becomes dissociated from the verb *had*, and takes on a syntactic life of its own. In other words, from a syntactic point of view, we might pose the question, where does the VP whose head is *had* in line 1 'end'?

Once again, there is nothing interactionally bizarre about what is happening in this conversation, and the answer to this syntactic puzzle doesn't seem to matter very much for the participants, who are doing a perfectly competent job of jointly manipulating the syntactic resources available to them to achieve a socially appropriate interaction.

---

[10] For valuable discussion of a range of interactional factors involved in the joint activity of listing, see (Jefferson 1990 and Erickson 1992).

Consider the following example, in which the dissociation from the original schema is more clearly observed:

(29) Carsales 5

```
 1 G:    .. (H) the only thing you can do is be the best you can.
 2       .. [right]?
 3 D:        [but definitely].
 4 G:    .. [2 that's it 2].
 5 D:    .. [2 and let her 2] know that,
 6 G:    .. yeah.
 7 D:    .. let her know that I still ca=re,
 8       .. an=d,
 9       I'm not getting invo=lved with anybody else.
10 G:    ...() yeah.
11 D:    .. because I don't have the time.
12       .. right now I have a caree=r.
13       I have goals set for myself,
14       also,
15       I want to make fifty thou a year,
```

D has been talking about the possibility of getting back together with his former girlfriend. In the discourse preceding this segment, D has admitted that the reason for their breaking up was his immaturity, which he is now trying to change by establishing his own career as a car salesman. In lines 7-9, D first says *let her know that I still care and I'm not getting involved with anybody else*. After a long pause and a continuer *yeah* from G (Schegloff 1982) in line 10, D continues what appears to be an instantiation of a *because*-clause schema, as in (30):

(30) CLAUSE because CLAUSE

Based on the syntax and the semantics of this clause combination, we could imagine that this *because*-clause gives a reason for D not getting involved with anybody else. But, interestingly, given what D is trying to convey, that interpretation is not possible. In other words, D cannot be saying that he wants to let her know that he still cares, and that he's not getting involved with anybody else *because he doesn't have the time*: that would not be what he wants her to think is the reason for his not getting involved with anyone else. That is, in terms of syntax, this *because*-clause in line 11 cannot possibly be part of the schema instantiated in lines 7–9, which involve the main clause *let her know*; this *because*-clause is NOT a part of the complement to the main clause *let her know*. Rather, it is understood as offering an account of the clause *I'm not getting involved with anybody else* in line 9, which now seems to have been 'dissociated' from the main clause *let her know* and is taken as an independent clause *directed to G* only as the recipient of the information:

I'm not getting involved with anybody else because I don't have the time.

This analysis is supported by the previous discourse context where D has been telling G about his past immature behavior, his attempt to change it by taking on a new career, his renewed feelings toward his former girlfriend, and his very busy new life. So we see here that an utterance produced as part of one schema has been redesigned as part of a new schema in line with the development of D's current local interaction with G.

Again, what these 'dissociation' examples show is that as interaction proceeds, we analysts may be presented with what can only be described in terms of a schema instantiation which seems to have 'lost' the schema with which it should have originally been associated. But this is precisely because syntax is locally managed: conversationalists are typically concerned with what is happening at a very local level, and may not be 'keeping track' of which syntactic productions instantiate which schemas.

3.2.3 'Messed up' Syntax: Final Example

In this section, as our final exploration of 'messed-up' syntax, we examine an example in which the interpretation of utterances crucially depends on the interaction; additions are made to propositions that are never uttered, but this does not appear to cause any problem for the participants:

> (31) Lam 8
> (talking about a pregnant neighbor of J and H)
> 1 H: does she even have a b- a man-?
> 2   I guess she must.
> 3 M: ...() does she have a what?
> 4 J: [a ma=n].
> 5 H: [a ma=n].
> 6 P: @@@
> 7 J: she has some [kind of a] —
> 8 M: [at least] temporarily,
> 9 P: [yeah],
> 10 H: [yeah].
> 11 J: @@@ (H) [at one time].
> 12 H: [for about five minutes],
> 13   probably.

This example illustrates several of the principles we have discussed in this paper, and touches on a new one. Clearly there is collaboration: H, M, and J all contribute to the joint construction of the expression of an attitude towards the hapless neighbor. But, interestingly, from a syntactic perspective, we can see that no syntactic unit is ever explicitly produced which could serve as the basis for the three additions made by M, J, and H in lines 8, 11, and 12. To see this, let's look at the sequence starting from H's question in line 1. This question contains a replacement of *man* for what was presumably going to be the more explicit term *boyfriend*. H then offers an answer to his own question *I guess she must*. In line 3, M initiates a repair sequence to clarify what it was that H wondered whether she had. Upon receiving the clarification, M then proceeds to add the adverbial phrase *at least temporarily* to something which has never been uttered, but which can be semantically patched together from H's question and his own answer to it *I guess she must [have a man]* to form a kind of 'semantic proposition'. At the same time, J says *she has some kind of a—*, which is never finished. P and H together offer reactive tokens *yeah*, and then J and H 'add' prepositional phrases, neither of which 'goes with' anything uttered up to this point: J's *at one time* and H's *for about five minutes, probably* are both appropriate only with clauses expressing past time. But both J's unfinished clause at line 7, *she has some kind of a*—and H's earlier question-answer *does she ...? I guess she must* are both in the present tense. Once again, from a syntactic and semantic point of view, what these speakers are doing does not constitute instantiations of any clausal constructional schemas, but there is no evidence to suggest that any of them find any trouble with the interaction. As Schegloff 1979 has suggested, syntactic and semantic needs may be, and often are, subordinated to interactional needs.

3.3 Summary

What do these examples suggest for the study of syntax? One answer is that they show how closely conversational participants monitor and are attuned to their own and each other's ongoing syntactic production, that they can as readily add to each other's schema instantiation as to their own, which in turn suggests that they attend to abstract schemas, of which actual utterances are instantiations. These examples thus provide further support for the view that a knowledge of schemas is part of what conversationalists bring to an interaction. But the most recent set of examples we considered show that adherence to schemas may readily be overridden by more pressing interactional concerns. That is, instances of overlap and dissociation reveal ways in which interactional considerations may produce instantiations which bear little relation to any schema.

Another answer is that they show that syntactic knowledge is a form of socially shared cognition (Saussure 1959, Schegloff 1991); what we *know*w-3 cannot be separated from what we *do*. We think then that examples such as those we have been discussing suggest that syntax can also be seen as an instance of "the inextricable intertwinedness of cognition and interaction," in which participants are jointly maintaining "a *world* (including the developing course of the interaction itself) mutually understood by the participants as some *same* world" (Schegloff 1991: 151–152). The schema instantiations which are started by one speaker and completed by another show that each speaker construes his/her contribution to be (part of) an instantiation of the *same* schema, and suggest that syntax cannot be thought of except as drawing upon both cognitive and social resources.

4. Conclusions

What we have tried to show in this paper is a set of instances from our conversational English database which illustrate some of the ways in which the interface between syntax and interaction is most evident. We have shown several frequently occurring types of situations in which the effect of interaction on syntax can clearly be investigated. Taking the production of syntactic units as a process of realizing constructional schemas, we have seen that speakers often cooperate in the production of such realizations, adding to their own or each other's realizations in ways that yield realizations that are in other contexts produced by a single speaker. We have considered a number of instances in which speakers syntactically collaborate in the interest of achieving successful interactions. These collaborations can be classified in syntactically interesting ways, in terms of whether syntactic units are finished or not, and who finishes or adds to what, as we have tried to show, but our main point should be clear: that we can learn a great deal about syntax by looking at syntax in real life: 'on-line' syntax as it is used by speakers engaged in actual, ordinary, everyday interaction. What this suggests is that not only conversations, but also syntax itself, must ultimately be understood as intersubjective and jointly constructed. Because speakers share knowledge of schemas and can recognize others' talk as contributions to their own syntactic productions, a model of syntax must include this rather remarkable ability to cooperate in the achievement of mutually satisfying interactions by means of shared syntactic resources.

Syntactic knowledge, then, cannot be just something static that speakers 'carry around in their heads', but must be understood in a much more dynamic way as a resource that guides the production and interpretation of utterances. At the same time syntactic knowledge is in no way the only thing which determines the form that utterances can take. Syntactic knowledge is constantly being modified by conversational encounters, and is drawn on in innovative ways to achieve satisfying interactions. And the 'bottom line' for interactants does not appear to be syntactic 'well-formedness', which can and often is foregone, but rather relevant interactional actions (as shown in more depth in Schegloff (1995)). We have seen how clauses and clause combinations are broken down when interactionally relevant actions are called for, thus confirming Schegloff's (1979:269) claim that:

> What is thought of in terms of current syntax and the 'integrity' of the sentence is, therefore, systematically subordinated to other sequential requirements.

It has been suggested that much of what we learn from examining the syntax of ordinary conversational data is that we do not speak in 'fully grammatical utterances'. But in our view, the point is not so much that we don't always speak in 'fully grammatical utterances', but that what it *means* to be 'fully grammatical' or 'not fully grammatical' needs to be re-examined in the light of the way grammar and interaction work together; the very nature of our grammatical resources and how we make use of them both depend to a very great extent on what interactional demands we are trying to satisfy.

Acknowledgments

We wish to acknowledge the work of John Du Bois on the Corpus of Spoken American English, and corpus project members, especially Danae Paolino, for the collection and transcription of much of the data for this study. We would also like to thank Eduard Hovy for comments on an earlier version of this paper.

Appendix: Symbols for Discourse Transcription

From Du Bois (1991) and Du Bois et al. (1993).

UNITS

Intonation unit	{carriage return}
Truncated intonation unit	—
Word	{space}
Truncated word	-

SPEAKERS

Speaker identity/turn start	:
Speech overlap	[]*

TRANSITIONAL CONTINUITY

Final	.
Continuing	,
Appeal	?

LENGTHENING   =

PAUSE

Long	...()
Medium	...
Short	..
Latching	(0)

VOCAL NOISES

Vocal noises	()

Alveolar click	(TSK)
Inhalation	(H)
Exhalation	(Hx)
Glottal stop	%
Laughter	@
Nasal laughter	@N

QUALITY

Quality	<Y Y>
Laugh quality	<@ @>
Quotation quality	<Q Q>
Forte; loud	<F F>
Higher pitch level	<H H>
Parenthetical prosody	<PAR PAR>
Whispered	<WH WH>
Creaky	<% %>

TRANSCRIBER'S PERSPECTIVE

Researcher's comment	(( ))
Uncertain hearing	<X X>
Indecipherable syllable	X

* Certain brackets are indexed with numbers to clarify which speech overlaps with which.

References

Ashby, W. 1988. The syntax, pragmatics, and sociolinguistics of left- and right-dislocations in French. *Lingua* 76, 203–229.

Ashby, W. 1992. The variable use of *on* versus *tu/vous* for indefinite reference in spoken French. *Journal of French Language Studies* 2, 135–157.

Barlow, M. and S. Kemmer. To appear. A schema-based approach to grammatical description. In R. Corrigan, G. Iverson, and S. Lima (eds.) *The Reality of Linguistic Rules*. Amsterdam: John Benjamins.

Bentivoglio, P. 1992. Linguistic correlations between subjects of one-argument verbs and subjects of more-than-one-argument verbs in spoken Spanish. In P. Hirschbühler and K. Koerner (eds.) *Romance Languages and Modern Linguistic Theory*. Amsterdam: John Benjamins. 11–24.

Chafe, W. 1980. The deployment of consciousness in the production of a narrative. In W. Chafe (ed.) *The Pear Stories: Cognitive, Cultural, and Linguistic Aspects of Narrative Production*. Norwood, NJ: Ablex, 9–50.

Chafe, W. 1987. Cognitive constraints on information flow. In R. Tomlin (ed.) *Coherence and Grounding in Discourse*. Amsterdam: Benjamins. 21–51.

Chafe, W. 1988. Linking intonation units in spoken English. In J. Haiman and S.A. Thompson (eds.) *Clause Combining in Discourse and Grammar*. Amsterdam: John Benjamins. 1–27.

Chafe, W. 1993. Prosodic and functional units of language. In J.A. Edwards and M.D. Lampert (eds.) *Talking Data: Transcription and Coding Methods for Language Research*. Hillsdale, NJ: Lawrence Erlbaum. 33–43.

Chafe, W. 1994. *Discourse, Consciousness, and Time: The Flow and Displacement of Conscious Experience in Speaking and Writing*. Chicago: University of Chicago Press.

Clark, H. and D. Wilkes-Gibbes. 1986. Referring as a collaborative process. *Cognition* 22, 1–39.

Croft, W. 1991. *Syntactic Categories and Grammatical Relations*. Chicago: University of Chicago Press.

Cruttenden, A. 1986. *Intonation*. Cambridge: Cambridge University Press.

Crystal, D. 1969. *Prosodic Systems and Intonation in English*. Cambridge: Cambridge University Press.

Du Bois, J. 1985. Competing motivations. In John Haiman (ed.) *Iconicity in Syntax*. Amsterdam: John Benjamins.

Du Bois, J. 1991. Transcription design principles for spoken discourse research. *Pragmatics* 1, 71–106.

Du Bois, J. and S. Schuetze-Coburn. 1993. Representing hierarchy: Constituent structure for discourse databases. In Jane A. Edwards and Martin D. Lampert (eds.) *Talking data: Transcription and Coding Methods for Language Research*. Hillsdale, NJ: Lawrence Erlbaum. 221–260.

Du Bois, J., S. Schuetze-Coburn, D. Paolino, and S. Cumming. 1993. Outline of discourse transcription. In J.A. Edwards and M.D. Lampert (eds.) *Talking data: Transcription and Coding Methods for Language Research*. Hillsdale, NJ: Lawrence Erlbaum. 45–89.

Duranti, A. 1994. *From Politics to Grammar*. Berkeley: University of California Press.

Duranti, A. and E. Ochs. 1979. Left-dislocation in Italian conversation. In T. Givón (ed.) *Discourse and Syntax*. New York: Academic Press. 377–416.

Erickson, F. 1992. They know all the lines: Rhythmic organization and contextualization in a conversational listing routine. In P. Auer and A. di Luzio (eds.) *The Contextualization of Language*. Amsterdam: John Benjamins. 365–397.

Ferrara, K. 1992. The interactive achievement of a sentence: Joint productions in therapeutic discourse. *Discourse Processes* 15, 207–228.

Ferrara, K. 1994. *Therapeutic Ways with Words*. Oxford University Press .

Fillmore, C.J. 1988. The mechanisms of 'construction grammar'. *Berkeley Linguistics Society* 14, 35–55.

Ford, C.E. 1993. *Grammar in Interaction: Adverbial Clauses in American English conversations*. Cambridge: Cambridge University Press.

Ford, C.E. and S.A. Thompson. To appear. Interactional units in conversation: Syntactic, intonational, and pragmatic resources for the projection of turn completion. In E. Ochs, E. Schegloff, and S.A. Thompson (eds.) *Grammar and Interaction*.

Fox, B.A. 1987. *Anaphora and the Structure of Discourse*. Cambridge: Cambridge University Press.

Fox, B.A. 1994. Contextualization, indexicality, and the distributed nature of grammar. *Language Sciences* 16, 1–37.

Fox, B.A. and R. Jasperson. To appear. The syntactic organization of repair. In P. Davis (ed.) *Descriptive and Theoretical Modes in the New Linguistics.*

Fox, B.A., R. Jasperson, and M. Hayashi. To appear. Resources and repair: A cross-linguistic study of the syntactic organization of repair. In E. Ochs, E. Schegloff, and S.A. Thompson (eds.) *Grammar and Interaction.*

Fox, B.A. and S.A. Thompson. 1990a. A discourse explanation of the grammar of relative clauses in English conversation. *Language* 66, 297–316.

Fox, B.A. and S.A. Thompson. 1990b. On formulating reference: An interactional approach to relative clauses in English conversation. *Papers in Pragmatics* 4, 183–195.

Geluykens, R. 1987. Tails (right-dislocations) as a repair mechanism in English conversation. In J. Nuyts and G. De Schutter (eds.) *Getting One's Words into Line: On Word Order and Functional Grammar.* Dordrecht: Foris. 119–129.

Geluykens, R. 1989. Referent-tracking and cooperation in conversation: Evidence from repair. *Papers from the 25th Regional Meeting of the Chicago Linguistic Society.* Chicago: Chicago Linguistic Society. 65–76.

Geluykens, R. 1992. *From Discourse Process to Grammatical Construction: On Left-Dislocation in English.* Amsterdam: John Benjamins.

Geluykens, R. To appear. The pragmatics of discourse anaphora in English: evidence from conversational repair.

Goodwin, C. 1979. The interactive construction of a sentence in natural conversation. In G. Psathas (ed.) *Everyday Language: Studies in Ethnomethodology.* New York: Irvington.

Goodwin, C. 1980. Restarts, pauses and the achievement of a state of mutual gaze at turn beginning. *Sociological Inquiry* 50, 272–302.

Goodwin, C. 1981. *Conversational Organization: Interaction Between Speakers and Hearers.* New York: Academic Press.

Goodwin, C. and M.H. Goodwin. 1987. Concurrent operations on talk: Notes on the interactive organization of assessments. *IPRA Papers in Pragmatics* 1.1, 1–54.

Goodwin, C. and M.H. Goodwin. 1992. Context, activity and participation. In P. Auer and A. di Luzo (eds.) *The Contextualization of Language.* Amsterdam: John Benjamins. 77–99.

Iwasaki, S. and H. Tao. 1993. A comparative study of the structure of the intonation unit in English, Japanese, and Mandarin Chinese. Paper presented at the annual meeting of the Linguistic Society of America, January 1993.

Jefferson, G. 1973. A case of precision timing in ordinary conversation: Overlapped tag-positioned address terms in closing sequences. *Semiotica* 9, 47–96.

Jefferson, G. 1990. List construction as a task and interactional resource. In G. Psathas (ed.) *Interactional Competence,* 63-92. Washington: University Press of America.

Jefferson, G. and E. Schegloff. 1975. Sketch: Some orderly aspects of overlap in natural conversation. Unpublished ms.

Keenan, E. and B. Schieffelin. 1976. Foregrounding referents: A reconsideration of left dislocation in discourse. *Berkeley Linguistics Society* 2, 240–257.

Lakoff, G. 1987. *Women, Fire, and Dangerous Things.* Chicago: University of Chicago Press.

Lambrecht, K. 1984. A pragmatic constraint on lexical subjects in spoken French. *Papers from the 20th Regional Meeting of the Chicago Linguistic Society,* 239-256. Chicago: Chicago Linguistic Society.

Lambrecht, K. 1986. Pragmatically motivated syntax: Presentational cleft constructions in spoken French. *Papers from the Parasession at the 21st Regional Meeting of the Chicago Linguistic Society*. Chicago: Chicago Linguistic Society. 115–126.

Lambrecht, K. 1987. On the status of SVO sentences in French discourse. In Russell S. Tomlin (ed.) *Coherence and Grounding in Discourse*. Amsterdam: John Benjamins. 217–261.

Langacker, R.W. 1987. *Foundations of Cognitive Grammar*, Vol. 1.   Stanford: Stanford University Press.

Langacker, R.W. 1991. *Foundations of Cognitive Grammar*, Vol. 2. Stanford: Stanford University Press.

Lerner, G.H. 1987. *Collaborative Turn Sequences: Sentence Construction and Social Action*. Unpublished Ph.D. dissertation, University of California, Irvine.

Lerner, G.H. 1989. Notes on overlap management in conversation: The case of delayed completion. *Western Journal of Speech Communication* 53, 167–177.

Lerner, G.H. 1991. On the syntax of sentences-in-progress. *Language in Society* 20, 441–458.

Lerner, G.H. In progress. Transforming 'dispreferreds' into 'preferreds': a systematic locus for preempting a turn at talk.

Lerner, G.H. To appear. On the 'semi-permeable' character of grammatical units in conversation: conditional entry into the turn space of another speaker. In E. Ochs, E. Schegloff, and S.A. Thompson (eds.) *Grammar and Interaction*. Cambridge: Cambridge University Press.

Maynard, S.K. 1989. *Japanese Conversation: Self-Contextualization through Structure and Interactional Management*. Norwood, NJ: Ablex.

Ochs, E., E. Schegloff, and S.A. Thompson (eds.) To appear. *Grammar and Interaction*. Cambridge: Cambridge University Press.

Ono, T. and R. Suzuki. 1992. Word order variability in Japanese conversation: Motivations and grammaticization. *Text* 12, 429–445.

Ono, T. and S.A. Thompson. 1994. Unattached NPs in English conversation. *Berkeley Linguistics Society* 20.

Ono, T. and S.A. Thompson. 1995a. What can conversation tell us about syntax? In P. Davis (ed.) *Descriptive and Theoretical Modes in the New Linguistics*. Amsterdam: John Benjamins. 29–45.

Ono, T. and S.A. Thompson. 1995b. The dynamic nature of conceptual structure building: Evidence from conversation. In A. Goldberg (ed.) *Conceptual Structure, Discourse and Language*. Center for the Study of Language and Information. Cambridge: Cambridge University Press. 105–139.

Ono, T. and E. Yoshida. 1996. A study of co-construction in Japanese: We don't "finish each other's sentences". In N. Akatsuka and S. Iwasaki (eds.) *Japanese/Korean Linguistics 5*. Stanford: CSLI, Stanford University. 201–245.

Östman, J-O. 1981. *You Know: A Discourse-Functional Approach*.   Amsterdam: John Benjamins.

Pawley, A. and F.H. Syder. 1977. The one clause at a time hypothesis. Unpublished.

Pawley, A. and F.H. Syder. 1983. Natural selection in syntax: Notes on adaptive variation and change in vernacular and literary grammar. *Journal of Pragmatics* 7, 551–579.

Sacks, H., E. Schegloff, and G. Jefferson. 1974. A simplest systematics for the organization of turn-taking for conversation. *Language* 50(4), 696–735.

Saussure, F. de. 1959. *Course in General Linguistics*. New York: Philosophical Library.

Schegloff, E. 1979. The relevance of repair to syntax-for-conversation. In T. Givón (ed.) *Discourse and Syntax*, 261-286. New York: Academic Press.

Schegloff, E. 1982. Discourse as an interactional achievement: Some uses of 'uh huh' and other things that come between sentences. In D. Tannen (ed.) *Analyzing Discourse: Text and Talk*. Georgetown: Georgetown University Press, 71-93.

Schegloff, E. 1987. Recycled turn beginnings. In G. Button and J.R.E. Lee (eds.) *Talk and Social Organization*. Clevedon, England: Multilingual Matters. 70-85.

Schegloff, E. 1988. Discourse as an interactional achievement II: An exercise in conversation analysis. In D. Tannen (ed.) *Linguistics in Context: Connecting Observation and Understanding*. Norwood, NJ: Ablex. 135-158.

Schegloff, E. 1991. Conversation analysis and socially shared cognition. In L. Resnick, J. Levine and S. Teasley (eds.) *Perspectives on Socially Shared Cognition*. Washington D.C.: American Psychological Association. 150-171.

Schegloff, E. 1995. Discourse as an interactional achievement III: the omnirelevance of action. *Research on Language and Social Interaction* 28, 201-233.

Schegloff, E. 1996. Turn organization: One direction for inquiry into grammar and interaction. In E. Ochs, E. Schegloff, and S.A. Thompson (eds.) *Grammar and Interaction*. Cambridge: Cambridge University Press. 52-133.

Schiffrin, D. 1987. *Discourse Markers*. Cambridge: Cambridge University Press.

Schuetze-Coburn, S. 1992. Prosodic phrase as a prototype. *Proceedings of the IRCS Workshop on Prosody in Natural Speech*, University of Pennsylvania, August, 1992. (Institute for Cognitive Research Report 92-37.)

Schuetze-Coburn, S. To appear. *Prosody, Grammar, and Discourse Pragmatics: Organizational Principles of Information Flow in German Conversational Narratives*. Amsterdam: John Benjamins.

Schuetze-Coburn, S., M. Shapley, and E.G. Weber. 1991. Units of intonation in discourse: acoustic and auditory analyses in contrast. *Language and Speech* 34, 207-234.

Tao, H. 1992. NP intonation units and referent identification. *Berkeley Linguistics Society* 18, 237-247.

Tao, H. 1993. *Prosodic, Grammatical, and Discourse Functional Units in Mandarin Conversation*. Ph.D. dissertation, UC Santa Barbara.

Thompson, S.A. and A. Mulac. 1991a. A quantitative perspective on the grammaticization of epistemic parentheticals in English. In Elizabeth Traugott and Bernd Heine, *Grammaticalization II*. Amsterdam: John Benjamins. 313-339.

Thompson, S.A. and A. Mulac. 1991b. The discourse conditions for the use of complementizer *that* in conversational English. *Journal of Pragmatics*- 3(15), 237-251.

Weber, E. 1993. *Varieties of Questions in English Conversation*. Amsterdam: Benjamins.

Wilson, T.P. and D.H. Zimmerman. 1986. The structure of silence between turns in two-party conversation. *Discourse Processes* 9, 375-390.

Wittgenstein, L. 1958. *Philosophical Investigations*. Edited by G.E.M. Anscombe and R. Rhees, translated by G.E.M. Anscombe, 2nd edition. Oxford: Blackwell.

Chapter 4
The Information Structure of the Sentence and the Coherence of Discourse

Eva Hajicová

Charles University, Prague

1. Introduction

In contemporary linguistics, discourse structure is analyzed mainly from the point of view of the coherence of discourse using coreference. In this paper, I attempt to substantiate the claim that for an adequate analysis of discourse from this perspective, it is necessary to pay due regard also to those aspects of discourse that rely heavily on an issue from the domain of sentence structure, namely the topic-focus articulation (TFA). I start with a short overview of the treatment of TFA in the Praguian functional description (Section 2) comparing it with some other treatments and devoting attention to some questions which are currently discussed in linguistic writings (Section 3). In Section 4, I indicate a possible approach to the analysis of discourse structure, taking as starting point the TFA of the utterances as component parts of the discourse. This approach is illustrated on a piece of text in Section 5.

2. Topic-Focus Articulation in the Functional Generative Description

In order to set my analysis into the formal framework we work with, I first give a very brief characterization of that framework; for a more detailed treatment and a more complex characterization of the basic notions we use, see especially Sgall et al. (1986); Hajicová and Sgall (1987); Hajicová (1994).

The (linguistic, literal) meaning of the sentence is represented in terms of an underlying structure (TR, tectogrammatical representation), which along with the syntax proper captures also TFA. (We work with dependency syntax, which has many advantages for the representation of TFA, but our observations about TFA are not bound to this particular syntactic theory.) With regard to TFA, it is necessary to distinguish in the TR's:

(i) *contextually bound* and *non-bound* nodes (the former are used as easily accessible to the hearer and thus often can be deleted in the surface or expressed by weak pronouns, the latter represent 'new' or irrecoverable information); in the primary case, contextually bound elements (*b*) belong to the topic and contextually non-bound elements (n) to the focus, but this is not necesarily the case in embedded structures. In the examples, % denotes the boundary between topic and focus:

(1) (How has John arranged his books in his apartment?)

(2) He$_t$ placed$_t$ the books$_t$ on nature$_t$ % in his$_t$ bedroom$_n$.

(ii) the *topic* part (T) and the *focus* part (F) of the sentence; before we discuss the issues of defining these two notions (in Section 3 below), we characterize them here by an illustration: the difference between the two shapes of the sentences in the following examples is expressed in Czech by the placement of the two relevant noun groups at the beginning and at the end of the sentence, respectively; this is also the case with (3) in English, but here the word order shift is often accompanied by a change in the grammatical construction (in (4) and (5)) or by a different placement of the intonation centre (in (6)):

(3) (a) I study linguistics on Sundays.

(b) On Sundays I study linguistics.

(4) I have one good piece of news and one bad piece of news. The good news is that the Czechs made the revolution. The bad is that the revolution was made by the Czechs. (English translation of a Czech text in a cartoon)

(5) The matter is not that Prof. Janouch bought the Leksell gamma knife, but that the Leksell gamma knife was bought by Professor Janouch. (English translation of a piece of Czech newspaper article)

(6) (a) STAFF behind the counter.

(b) staff behind the COUNTER.

(iii) *deep word order* (the hierarchy of communicative dynamism): in both (7)(a) and (b), the wh-clause is in T, yet the order of the two noun groups in T is semantically relevant in the same way as the order of these two noun groups in example (8):

(7) (a) It was JOHN who talked to few girls about many problems.

(b) It was JOHN who talked about many problems to few girls.

(8) (a) John talked to few girls about many problems.

(b) John talked about many problems to few girls.

On the preferred readings of both the (a) sentences, there was a small group of girls to which John talked about many problems (i.e., to the same group), whereas in the (b) sentences the preferred readings concern a single set of (many) problems.

In (7) and (8) the deep word order of the *to-* and *about*-clauses is identical to their surface order. However, the position of *John* in the surface order differs from its placement in the deep word order: the position of the intonation center shows that *John* is the most dynamic element of the sentence, i.e., placed rightmost in the deep word order.

3. Some Open Questions

Even though it might seem that TFA has received much attention in the context of European and American linguistics, several important questions still deserve further elaboration and discussion.

One of the points is the question of which level of the language system TFA belongs to. In some frameworks the phenomena of TFA are included in the realm of pragmatics; this is, for example, the approach of Dik's functional grammar, where sentences differing only in their topic/focus articulation are understood to correspond to a single proposition (cf. Gebruers 1983), or the approach of several linguists in their treatment of negation, e.g., Horn (1989), Atlas (1977), and to a certain extent also Wilson (1975) and Kempson (1975). Others, in contrast, treat TFA as a phenomenon of the domain of discourse rather than of sentence structure.

Our approach takes a different standpoint: TFA, together with the scale of communicative dynamism (corresponding to the deep word order, essentially determined by a combination of surface order and stress pattern), is regarded as a *semantically* relevant distinction, the place of which is therefore on the level of the linguistic (literal) meaning of the sentence, i.e., of (underlying) syntactic structure (having a similar interface role as Chomsky's logical form): a sentence with ambiguous topic/focus articulation has at least as many representations on this underlying level as the number of articulations of its topic and focus (and as many deep word orders) that can be assigned to it. The thesis about the necessity of accounting for topic and focus on the underlying level was put forward by Sgall (1967), where he also shows that TFA belongs to the grammar of language; he illustrates this on the difference between the weak and strong forms of personal pronouns in Czech (e.g., *ho* vs. *jeho* in Acc. Sing. of *on* [he]), where the weak form is used with contextually bound items. The Japanese particle *wa* provides additional grammatical evidence.

The semantic interpretation of negation brings another supporting argument for such a standpoint. Investigations described by Hajicová (1973) have led to the conclusion that, in negative sentences, the scope of negation prototypically is identical to the focus part of the sentence:

(9) (a) Jim didn't come to the PARTY (, since Jane was ill).
    (verb included in the focus)

   (b) Jim didn't come TODAY (, but he was here yesterday).
    (verb in the topic, not negated)

In non-prototypical cases, either only a part of the topic is negated, see (10), or the operator of negation itself constitutes the whole focus, and the topic represents its scope, as in (11):

(10) Jim didn't come to the party since Jane was ILL.
(e.g., answering: Why didn't Jim come to the party?)

(11) Jim DIDN'T come to the party.

The behaviour of other 'focalizers' (focus-sensitive operators), such as *even, only, also*, and at least some sentential adverbials (cf. Koktová, 1986), is similar to that of negation:

(12)(a) Jim only came to the PARTY (and didn't take part in the excursion).
(verb included in the focus, cf. (9)(a))

(b) Jim came only to the PARTY (and not to the theatre performance).
(verb in the topic, cf. (9)(b))

(13) (Why did Jim come only to the party, not to the performance?) Jim came only to the party since Jane was ILL.
(the verb and focalizer in the topic, cf. (10))

(14) Jim ONLY came to the party. (at least on one reading, the focalizer itself constitutes the whole focus, cf. (11))

Rather than working with a single dichotomy, some linguists prefer to speak about additional layers. Mathesius (1939) proposes to distinguish between "the starting point" of the utterance and the "theme". A twofold division is best known from Halliday's (1967) distinction between information structure ('given' versus 'new') and thematic structure. Also in some more recent writings the authors propose multi-layered treatments, as for instance Jacobs (1983) when analyzing the so-called Gradpartikeln (focalizers).

We have not found any substantial arguments for such a treatment. In Halliday's characterization of thematic structure, the theme is supposed to be that element of the sentence which is the first lexical item—a "hook" on which the utterance hangs, as Halliday puts it. If this is so, i.e., if the term *theme* is taken as a label for such a positionally determined item, then the necessity of this distinction is undermined by the fact that the first lexical position of the sentence can be occupied by elements serving for very different functions, see the examples (15) through (18) with capital letters denoting a marked position of the intonation center):

(15) (Yesterday, I visited my mother.) My mother is a remarkable person.

(16) (What has happened that everybody is so excited?) KENNEDY was assassinated.

(17) (It was) JOHN (who) missed the train yesterday.

(18) (Who was visited also by John?) Only JIM was visited also by John.

Vallduví (1990) in his approach to 'information packaging' (a framework based on Heim's (1982) file change semantics and Prince's (1985) understanding of the stock of knowledge) also finds it useful to work with a more structured treatment, distinguishing within his 'ground' (comparable to our notion of topic) between a 'link' (L) and a 'tail' (T):

(19) The boss HATES broccoli.

      L   T

Again, we assume that this distinction (which should be viewed independently of the constraints on word order in English) can be accounted for by means of the scale of communicative dynamism (CD): the link (close to Lötscher's, 1983, 'new topic') seems prototypically to correspond to the element with the lowest degree of CD (what we call 'topic proper'), and the tail to the element(s) with a degree of CD higher than that of topic proper; the notion 'focus' has the same extension in the two frameworks.

Firbas, who introduced the notion of communicative dynamism into the studies of topic and focus, defined topic (theme) as the communicatively least important element(s) of the sentence (Firbas 1964). This definition invokes a number of questions, the most striking being connected with the existence of what we understand as sentences without a topic, such as a typical sentence that brings "hot news" (e.g., (16) above) or sentences opening the discourse (of course, not necessarily always sentences without topic). In such sentences, there is a hierarchy of CD, therefore there is a communicatively least important item but this item is part of the focus, of the information presented as "new" (in his most recent work, however, Firbas (1992:73) admits the existence of sentences without theme).

Within the formal system as characterized in Section 2 above, Sgall (1979) presented a formal definition of focus, based on the primitive notion of contextual boundness (CB) and the representation of the underlying structure of the sentence (TR—tectogrammatical representation) as a dependency tree. Focus then can be defined in three steps:

(i) If the main verb of the TR or some of the nodes which directly depend on it are NB, then these nodes belong to the focus of the TR.

(ii) If a node other than the root belongs to the focus, then all nodes subordinated to it also belong to the focus.

(iii) If the root and all of its daughter nodes are CB, then it is necessary to specify the rightmost daughter node and look for its NB daughter nodes; if all are CB, then look for the NB daughter nodes of this set of nodes, and so on. These NB nodes constitute the focus, together with nodes specified under (ii).

Those parts of the sentence (more precisely, of its TR) which do not belong to its focus are then defined as constituting its topic.

Computational models of aspects of discourse have often been constructed in close connection with research in the experimental domain of Artificial Intelligence; an interesting approach to the structure of discourse was presented by Grosz and Sidner (1986), following up the investigations of task dialogues by Grosz (1977). Their attentional structure of utterances closely corresponds to the TFA analysis, with one proviso: what is called *focus of attention* there, is the element *just introduced* by the speaker, who 'focuses' his/her attention on it.

Taking (20) as a part of discourse consisting of (a) and (b), then in terms of the AI-oriented research the focus of (20)(b) refers to the baby, since the baby is one of the items 'just introduced' (namely, by (20)(a)). In terms of linguistic analysis, however, the expression referring to the baby in (20)(b), namely the pronoun *it*, belongs to the topic rather than to the focus of this utterance; the (linguistic) focus of (20)(b) is *had been crying nearly all day.*

(20)(a) The mother picked up the baby.

   (b) It had been crying nearly all day.

The means of expression of TFA range from morphological devices (strong and weak pronouns, the topic particles of Japanese, Tagalog, etc.) to syntactic constructions (active-passive or cleft constructions, inversion verb forms such as *make out of* vs. *make into*, etc. in English, word order in Slavic languages, but to a certain extent also in English, cf. examples (3), (7), (8)) as well as prosodic phenomena (the position of intonation centre and intonation contours, whose relationships to focus have not yet been systematically described).

Two otherwise identical utterances differing only in the position of the intonation centre are two different sentences, rather than two variants of the same sentence. This holds for the example (6) above, where the (a) and (b) sentences clearly differ semantically. This is similar in (21) and (22):

(21)(a) Dogs must be CARRIED.

   (b) DOGS must be carried.

(22)(a) I do linguistics on SUNDAYS.

   (b) I do LINGUISTICS on Sundays.

   (c) On Sundays, I do LINGUISTICS. (= b)

In all these examples, the bearer of the intonation centre is the most dynamic element of the sentence, i.e., its focus proper, although it may occupy a non-prototypical position (not at the end of the sentence). These are the most evident cases of differences between the scale of CD and surface word order. Other such asymmetries are present within noun groups (with an adjective to the left and a prepositional group to the right of the noun, even if this does not correspond to CD), or, in many languages, the postions of the clitics, of the verb, and so on.

4. TFA and Discourse Patterns

According to the mutual relations of TFA of individual utterances in the discourse, several strategies of the speaker/writer can be recognized: Weil (1944) establishes a distinction between 'marche parallèle' and 'progression', Mathesius (1942) evaluates the links between topics and foci of individual utterances from the stylistic point of view, Daneš (1970) recognizes four types of 'thematic progression' in the text.

At every time point of a discourse, the speaker assumes to share with the hearer certain pieces of information (the stock of shared knowledge, SSK). The SSK is not just a collection of pieces of information (referred to by expressions) but also specifies a certain ordering of their salience (prominence, activation). In the prototypical case, the speaker chooses some of the activated (established) items with which to start his or her utterance (constituting the topic), and then developing them further, by modifying, changing or adding to these pieces of information according to that part of the stock which he or she assumes not to be known (or easily accessible) to the hearer. If the development of the discourse, utterance after utterance, is viewed from this point, the coherence of the discourse with its anaphorical relations can be analyzed in terms of the changes in activation of the items of SSK in close connection with the TFA of individual utterances.

An interesting question that arises in this connection is to what extent the TFA of individual sentences is relevant for the changes of the degrees of salience (activation, prominence), which help the hearer to identify the reference of referring expressions. The following five observations are pertinent:

(a) It is rather obvious that, at every time point of the discourse, the items referred to by the (parts of the) focus of the immediately preceding utterance are the most activated ones.

(b) If an item is referred to in the topic part of the sentence, then at least two issues are to be taken into consideration:

(i) A pronominal reference strengthens the activation of the item referred to to a lesser degree than a reference using a full (definite) NP.

(ii) The activation of the items referred to in the topic part of a sentence fades away less quickly than that of the items referred to only in the focus part.

(c) If the degree of activation of an item A is changed (lowered or raised), then also the degree of activation of the items associated with the object referred to by A (by an is-a relation, by a part-of relation, etc.) is changed in the corresponding direction. Also other scales or hierarchies should be considered, such as a more or less immediate associative relationship, or that of prominence with regard to the individual utterances and their positions in the text (e.g., sentences metatextually opening a narration or one of its portions, headings, etc., are more prominent than other elements of the text, in that the objects introduced there display a relatively high degree of salience).

(d) If an item of the stock of the shared knowledge is neither referred to in the given utterance, nor included among the associated objects, then its activation decreases; as mentioned in (b)(ii) above, the drop in activation is quicker if the item was referred to in the focus of the preceding utterance, and slower if it was referred to in the topic part.

(e) Certain specific expressions in particular languages give prominence to the items they precede by raising their activation more than otherwise would be the case. This holds, e.g., for the English phrases *as for...* and *concerning...*, that function as 'thematisers' and introduce a topic that was not expected; usually then the utterance recalls an object that is being mentioned only after an occurrence of several intervening utterances.

Representing discourse flow in this way (for a more detailed discussion and an example, see Hajicová 1987), a certain clear patterning can be observed, segmenting the discourse more or less distinctly into smaller units according to which items are most activated in these stretches. Also when searching for the 'topics of the discourse', one can look for items highly activated for the whole discourse or for some of its parts.

5. An Illustration

The analysis of a living text throws interesting light on the issues related to the discourse patterning as characterized above. For an illustration we have chosen a piece of (a slightly modified, mainly by shortening) text from the weekly news magazine *Time*. The passages relevant to our discussion are referred to by numbers occurring in parentheses before the corresponding segments. The expressions referring in the text to the items of knowledge (the salience of which is discussed below) are marked by italics.

Presidential Invasion

(1) On the afternoon of his 47th birthday, seven months after he took the oath of office, the President came to rest on a New England island Martha's Vineyard. (2) The island is so small that it has no traffic lights. (3) Its inhabitants are sufficiently celebrity-trained. (4) No one stares into opera diva Beverly Sills' grocery cart at Cronig's or gawks at Jackie Onassis riding her bike near her house in Gay Head. (5) A live-and-let-live attitude of the inhabitants of the island toward the famous is one reason of the President's decision. (6) Clinton is a man without a country house. (7) He doesn't even take off weekends, (8) and he delayed making holiday plans for a long time. (9) It was clear that Martha's Vineyard might be perfect for the Clintons, (10) but there was some apprehension that the First Vacationers would not be perfect for Martha's Vineyard. (11) It seemed that its tiny community is so set in their reverse-chic ways (23) that the newcomer cannot hope to adapt. (13) The celebrities, however, gave an important party for the President's birthday. (14) It was a jolly, casual evening. (15) The party didn't break up until 1 a.m., (16) unheard of on Martha's Vineyard, where, as Beverly Sills puts it, 10 p.m. is midnight.

(*Time*, August 30, 1993, Margaret Carlson: Presidential Invasion, pp. 30–31; abridged and modified)

As mentioned above, in a coherent discourse, only an item contained in the salient part of the stock of knowledge and thus assumed by the speaker to be shared by him/her and the hearer may be referred to in the topic part of the sentence. This holds for *the President* and *he* in (1), as well as for the *birthday* in the same utterance: the mental image of the President is installed in the heading of the article and the possessive pronoun *his* makes the birthday a part of the established portion of the SSK. *Inhabitants* in (3) illustrates the same point: the island has been introduced in (1) into the SSK, has received a high degree of salience by being mentioned in the focus part, and this degree has been retained by mentioning the island in the topic part of (2) and also by referring to it by the possessive pronoun *its* in (3). The inhabitants of that island as an associated item (cf. (c) above) thus are sufficiently present in the foreground to be referred to in the topic part of the utterance.

The rest of the items are introduced first in the focus part of the utterances: besides the island itself, as we have just mentioned, this is true about the traffic lights in (2), the celebrities in (3), the President's decision in (5), the country house in (6), the weekends in (7), and the birthday party in (13). As such, all these items are activated by their mentioning in the focus part to the highest degree (cf. (a) above).

If these items are not mentioned in the subsequent utterances, their activation fades away rather quickly (cf. (b)(ii) above); this is the case with the traffic lights, the country house and the weekends. These items never come back to the foreground. The salience of an item is preserved if the item is rementioned in the subsequent utterance; the fading of such an item during the subsequent utterances in which it is not mentioned then is slower.

An item can be reintroduced into the discourse in the topic of an utterance if its fading has not reached a certain threshold, which is the case of the reference to the President's decision: it is first mentioned in the focus of (5) and then referred to as *holiday plans* in the topic of (8). The same holds true about the President (introduced in (1), rementioned as a contextually bound item in an embedded adjunct within the focus of (5) and then in the topic of (6)), about the Vineyard (fading after (5), and reappearing in the topic of (9)), and about the celebrities (fading after (5), referred to as *its tiny community* in the topic of (11)).

However, if the fading of an item goes beyond such a threshold, its reappearance can be achieved only by a reference in the focus part of the utterance again (as contextually non-bound), as is the case with the President's birthday: after having been mentioned in (1), the next reference is not made until (13), where this item has to be reintroduced in the focus part.

The structure of (9) and (10) is an interesting example how the choice of topic and focus can be used for subtle options concerning discourse strategy: both the President and the island are salient enough after the utterance of (8) that either of them can be used in the topic part of the next following utterance. The President was referred to by a pronoun in the topic part of (8) and the island in the topic part of (5), after being at the highest level of salience for all the five preceding utterances, so that the fading away of this item is very slow. The point under discussion in both (9) and (10) is the relation between the President and the (inhabitants of the) island. The writer then chooses one direction of the relation— that of the President to the island, and refers to the President in the focus part and to the island in the topic part of (9), while in (10), the order is reversed.

6. Concluding Remarks

We have concentrated in our discussion on one particular aspect of the analysis of discourse, namely on the impact of the TFA of individual utterances on the information flow of the discourse and on the changes of the degrees of salience of the items of SSK referred to in the given discourse. The approach sketched in Section 4 offers several more insights into the structure of discourse not touched here, one of them being the overall scheme of the information flow and the role of individual items of SSK in it: such a look would show that in our text, there are some items that occur in it only marginally, though being rather salient at some points of the discourse, while others stay in the foreground for a while but then (slowly) fade away and do not reappear (this is the case of the inhabitants of the island: being in the foreground during the utterances of (3) through (5), they do not come back again), or are permanently (after having been introduced into the discourse) above a certain threshold of salience (the island, the President, the celebrities).

We are convinced that this approach allows for a description of the dynamic aspects of discourse without losing track of the information structure of the individual component parts which are the building stones of the discourse as a whole. If topic and focus are understood as two parts of the (underlying) structure of the sentence, then analyses of the relationships between grammatical relations and the flow of information may gain a relatively reliable background. Among the objectives of further research there then belong issues such as the functional aspects of prosody, the possible positions of focalizers, contrastive topic, and discourse theme development with different ways of introducing a new topic, re-introducing a not-quite-recently-mentioned topic, or introducing an unexpected topic.

References

Atlas, J.D. 1980. Pragmatic and semantic aspects of negation. Linguistics and Philosophy 3, 411–414.

Daneš, F. 1970. Zur linguistischen Analyse der Textstruktur. Folia Linguistica IV, 72–78.

Firbas, J. 1964. On defining the theme in functional sentence analysis. Travaux Linguistiques de Prague 1, 267–280.

Firbas, J. 1992. Functional Sentence Perspective in Written and Spoken Communication. Cambridge: Cambridge University Press.

Gebruers, R. 1983. S.C. Diks Funktionale Grammatik: Eine Wallfahrt nach Prag? Linguistische Berichte 86, 49–67.

Grosz, B.J. 1977. The representation and use of focus in dialogue understanding. Technical Report 151, Artificial Intelligence Center, SRI International, Menlo Park, California.

Grosz, B.J., and C.L. Sidner. 1986. Attention, intentions, and the structure of discourse. Computational Linguistics 12, 175–204.

Hajicová, E. 1973. Negation and topic vs. comment. Philologica Pragensia 16, 81–93.

Hajicová, E. 1987. Focussing—A meeting point of Linguistics and Artificial Intelligence. In Ph. Jorrand and V. Zgurev (eds.) *Artificial Intelligence 2, Methodology, Systems, Applications.* Amsterdam: North Holland. 311–321.

Hajicová, E., 1994. Topic/Focus and Related Research. In P. Luelsdorff, (ed.) *Introduction to Praguian Linguistics.* Amsterdam: John Benjamins. 249–279.

Hajicová, E., and P. Sgall. 1987. The Ordering Principle. *Journal of Pragmatics* 11, 435–454.

Halliday, M.A.K. 1967. Notes on transitivity and theme in English. *Journal of Linguistics* 3 (1967) 37-81, 199–244; 4 (1968) 179–215.

Heim, I. 1982. *The Semantics of Definite and Indefinite Noun Phrases.* Ph.D. dissertation, University of Massachusetts, Amherst.

Horn, L.R. 1989. *A Natural History of Negation.* Chicago and London: The University of Chicago Press.

Jacobs, J. 1983. *Fokus und Skalen: Zur Syntax und Semantik der Gradpartikeln im Deutschen.* Tübingen: Niemeyer.

Kempson, R.M. 1975. *Presupposition and Delimitation of Semantics.* Cambridge: Cambridge University Press.

Koktová, E. 1986. *Sentence Adverbials in a Functional Description.* Amsterdam: John Benjamins.

Lötscher, A. 1983. *Satzakzent und funktionale Satzperspektive im Deutschen.* Tübingen: G. Narr.

Mathesius, V. 1939. O takzvaném aktuálním clenení vety (On the so-called functional sentence perspective). *Slovo a slovesnost* 5: 171–174. Translated as: On information-bearing structure of the sentence, in S. Kuno (ed.), *Harvard Studies in Syntax and Semantics,* 1975, 467–480.

Mathesius, V. 1942. Rec a sloh (Speech and style). In *Cteni o jazyce a poesii* (Readings on language and poetry). Prague: Melantrich, 11–102.

Prince, E.F. 1985. Fancy syntax and 'shared knowledge'. *Journal of Pragmatics* 9, 65–81.

Sgall, P. 1967. Functional sentence perspective in a generative description. *Prague Studies in Mathematical Linguistics* 2, 203–225.

Sgall, P. 1979. Towards a definition of focus and topic. *Prague Bulletin of Mathematical Linguistics* 31 (1979) 3–25; 32 (1980) 24–32. Printed in *Prague Studies in Mathematical Linguistics* 7 (1981), 173–198.

Sgall, P., Hajicová, E., and J. Panevova. 1986. *The Meaning of the Sentence in its Semantic and Pragmatic Aspects.* Dordrecht: Reidel and Prague: Academia.

Vallduví, E. 1990. *The Information Component.* Ph.D. dissertation, University of Pennsylvania.

Weil, H. 1844. *De l'Ordre des Mots dans les Langues Anciennes Comparées aux Langues Modernes.* Paris.

Wilson, D. 1975. *Presupposition and Non-Truth-Conditional Semantics.* New York.

Perspectives from
Computational Linguistics

Chapter 5
Discourse Coherence and Segmentation

Kathleen Dahlgren

Intelligent Text Processing Inc., Santa Monica

1. Introduction

This paper explores the basis of discourse structure, cognitive mechanisms for recovering it, and computational algorithms designed to mimic human discourse structure recovery for text. We argue that structure is recovered to the extent that the reader can build a coherent cognitive model of the eventuality (situation) the discourse describes from the reader's interpretation of the semantic content of the discourse. In empirical studies of newspaper commentary and narrative text we found that discourse structure is infrequently marked by cue phrases, and that paragraph shift, tense shift and focus shift do not add up to sufficient information for the location and recovery of discourse segment boundaries. A discourse theory which would rely solely upon these elements could not account for intuitions of discourse structure. In contrast, an adequate theory accounts for discourse in terms of coherence in addition to the above-mentioned elements. Asher (1993) and Asher and Kamp (1995) develop a similar theory.

This paper explores the basis of coherence relations and finds that they are explained by naive theories of causal and other structure in the world. The paper builds up a definition of discourse structure, and a method of segmenting discourses. We argue that discourse structure operates upon the content of the discourse, rather than upon the text or presentation, so that the theory structures sets of discourse entities in the semantics of the discourse. Discourse entities are abstract types (events, states, propositions) in the Discourse Representation Structure (Kamp 1981, Kamp and Reyle 1993, Asher 1993), a logical form for the discourse. In structured genres, the discourse structure is a tree with segments at the nodes, the global topic segment at the root, and arcs labeled by coherence relations. The representation captures both the explicit and implicit structure. The proposed theory of segmentation is validated by the constraints it places upon the location of antecedents of pronominal and other anaphoric devices in discourse structure trees.

Study 1

In a study of 16,000 words of *Wall Street Journal* (WSJ) text, the author examined global coherence (or segmentation) to determine which factors influenced it. For each new sentence, the possibility of a new sister- or sub-segment arises. Change of coherence relation was the most reliable indicator of a new segment and change of subject next most reliable. Other factors were paragraph indentation, length of segment with the same coherence relation to some other, cue phrases, and event anaphors. Significantly, there was a segmenting cue phrase such as, "turning to..." in only 16% of the cases of a new sister segment. Clearly any computational system which looks only for direct cues will miss most of the structure. Substantive coherence relations, if they can be extracted, are powerful indicators of discourse structure. This is study more fully described in Section 3.7.

Study 2

Another study of the same 16,000 plus 4,500 more words of *Wall Street Journal* text examined personal pronouns, demonstratives, and definite NP anaphora. We found that when the text in the earlier study is segmented, the resulting structure predicts constraints on anaphora resolution. This work empirically supports Grosz and Sidner (1986) and also shows that event and other abstract type anaphora has very different constraints from individual anaphora.

2. What Is Coherence?

The notion of coherence is based upon the intuition that the discourse in (1) seems to "hang together", while that in (2) does not. Empirically, to the extent that discourses do not cohere, they are difficult to interpret and remember (cf. Hobbs 1979, Van Dijk and Kintsch 1983, Graesser 1981).

(1) John invested heavily. He profited handsomely.

(2) John invested heavily. He ate pizza.

The coherence of (1) is said to be explained by the reader's inference of a goal relationship between the investing and the profiting (cf. Hobbs 1979, Hirst 1981, Fox 1984, Mann and Thompson 1987, Van Dijk and Kintsch 1983). Beyond this intuition there is little agreement as to a definition of coherence. We attempt to establish in this paper that a coherent discourse is one for which the hearer can build a cognitive representation such that the relations among individuals, events, states and other abstract types in the representation correspond with her/his understanding (naive theory) of the way actual world individuals and events relate. This is true even in the interpretation of metaphors and fiction. This is our fundamental point. It is extensible to account for hearer theories concerning mental states of interlocutors, coherent sets of propositions, logical arguments and so on. Although the representation may contain a variety of types of sensory images, in general, the hearer's "understanding" amounts to a naive theory (in the sense of Hayes 1985) of the causal and other structure of objects and events of Graesser (1981) or Van Dijk and Kintsch (1983). For (1), the hearer brings the discourse into accord with his/her theory about investing. We say 'theory' because very often people, and cultures, are quite mistaken in such causal inferences. Similarly, the world coheres under many conditions, but we humans may not understand. We may see some situation as chaotic, and only later discover its structure. Despite their incorrectness, people do use such structuring theories to manage the environment and to communicate via language. Because members of a subculture *share* naive theories, the speaker can juxtapose just the two sentences in (1), and know that the hearer will guess or infer that John's goal had been profit.

In summary, our view of coherence is that speakers in a given genre make a discourse (and thereby their reporting of events) understandable by choosing to report events using certain verbs in a certain sequence. This choice in a well-structured discourse makes it maximally possible for the hearer to build a cognitive model of these events which coheres. It will cohere to the extent that he/she can bring it into accord with her/his theories about the way the world works. The relationships in the naive theory of the world are causal, intentional, comparative, part-whole, etc.

Coherence is a gradient phenomenon in that the better-structured the discourse, the more readily and reliably a hearer will make coherence inferences. It is also genre-relative. Some conversations and literary styles are minimally coherent or deliberately incoherent. Similarly, the more knowledgeable and tuned in the hearer, the more accurately he/she will recover the speaker's intended cognitive representation.

Based upon the empirical study of coherence relations described below, a review of the literature on coherence relations, and an examination of the psycholinguistic and linguistic evidence for coherence relations, we settled upon seven coherence relations for which there was convergence from all sources (Dahlgren 1992). These are BACKGROUND, GOAL, CAUSE, CONSTITUENCY, CONTRAST, ELABORATION, and EVALUATION.

2.1 What Are Coherence Relations?

In this section the phenomenon of coherence is related to epistemology and psychology, in the hope that the terms of the theory may eventually be explained by psychological, sociological, and other human processes. Without a firm philosophical basis, we have no theory at all, but rather, a set of researcher intuitions. For example, a coherence relation such as CAUSE has been identified in several theories: presentation style (Hobbs 1985), pragmatics (Mann and Thompson 1987, hereafter M&T), and causal inferencing (van Dijk and Kintsch 1983, hereafter VD&K, Graesser and Clark 1985). The first question is, what coheres? Does discourse itself cohere (G&S, Hobbs), or the discourse situation (M&T), or a mental model of the content of the discourse (VD&K)? We argue that what coheres is a mental model of the situation a discourse describes, and that the way the mental model coheres is explained by naive theories of causal and other structure in the actual world. (Physical text structure marked by outlining or terms such "above" are not included in this discussion.)

What then is the coherence relation CAUSE? Following Davidson (1967b), we argue that it is a predication *cause(e1,e2)*, stated or implied in a text, which represents a speaker/hearer theory about the causal connections between events. Starting with the basics, what is an event (denotationally)? One standard view, with many problems, is that an event is a spatially and temporally located occurrence in the real world. A more sophisticated view of events is as concrete particulars individuated by their causes and consequences (Davidson 1967b). However, in terms of the cognitive phenomena of events and their structure in discourse, we must explain not the actual world itself, but rather human interpretation of the world, i.e., naive (meta)physics. Even the direct observation of some real event involves observer interpretation, minimally, of the idea that each frame or pixel he or she sees is part of the same real event and not a series of different events. The observer's view that something is happening, e.g., that car is falling off of a cliff, is a theory—often a conscious verbal theory. "Oh, that car is falling off of that cliff". When the observer thinks, "Oh, an accident", we have even more theorizing and interpretation as actor goals (or lack of them) are inferred. The predication of events is epistemological, heuristic, and retractable.

Now consider observer's interpretation of events. When a person has seen a car fall over a cliff, he/she may interpret it as a whole with two causally related subparts, as in (3).

(3) (e1) The car came to the edge.

(e2) The car fell over the edge.

The car came too close to the edge, a critical point was reached in which the car was no longer balanced on the edge, and it fell over.

How shall we analyze the observer's construction of causal inferences? In the cognitive interpretation of events (and of texts reporting events), people actively theorize and hypothesize about the causal structure of events as they unfold (Graesser 1981, Trabasso, Secco and van den Broek, 1983, Graesser and Clark, 1985, Trabrasso and Sperry, 1985). People are tentative about such inferences, but in order to function, they must guess. Such guesses are naive theories about the causal structure of events. Furthermore, the inferred causal structure is an important correlate of memory for the events. In example (3), such a naive hypothesis would be formed when the observer uses naive physics to figure out that (e1) and (e2) are causally related. In understanding, interpreting, and labeling observations of real world events like (3), members of a Western culture (and probably of any culture) infer that the critical point was reached, and that the coming close to the edge (in the end) caused the falling over the edge. The same holds for discourse and text understanding. As they read, people infer causal structure.

What, then, is the coherence relation CAUSE? It is an element of naive metaphysics, of naive theories about the relations between events. Next, in terms of semantic theory, how shall we handle the logical form of a coherence relation such as CAUSE? Davidson (1967a) argues against treating it as a logical connective like the material conditional, and in favor of treating it as a predicate. He considers sentences such as (4) and argues for a logical form like (5) .

(4) Jack fell down and broke his crown.

(5) $\exists$ e1,e2 (e1=*fall_down$_i$* $\wedge$ *e2=break_his_crown$_j$* $\wedge$ *cause(e1,e2)*))

In Davidson's analysis, a cause statement introduces a causal predication over events. He points out that the truth conditions of causal statements require that very many of the conditions present in a causal situation are part of the evaluation of truth, not just the descriptive predications for the events.

> ...we must distinguish firmly between causes and the features we hit on for describing them, and hence between the question whether a statement says truly that one event caused another and the further question whether the events are characterized in such a way that we can deduce, or otherwise infer, from laws or other causal lore, that the relation was causal. "The cause of this match's lighting is that it was struck.—Yes, but that was only *part* of the cause; it had to be a dry match, there had to be adequate oxygen, and the striking was hard enough, etc."

Davidson's explanation of explicit causal connectives can be extended to implicit (inferred) coherence relations. Then we interpret CAUSE relations as predications of events added to the semantic representation for a discourse, either because of direct surface forms like *cause* or *because*, or by hearer inference. Generalizing, then, the definition of coherence relations is:

> A coherence relation is part of a naive theory of the relation between elements introduced into a discourse. It is a binary predicate whose arguments are discourse individuals, discourse events or states, facts, propositions, event types, or sums of these (cf. Asher 1993). A coherence theory (predication) arises from naive theories about the causal and other structure of the world.

This definition refers to coherence inferences relating the contributions of two clauses of discourse. Relations among larger segments of discourse is the subject of Section 3. In brief, we will adopt the view of Asher (1993) that the relations which obtain between individual discourse events also obtain between structures of discourse events. Just as we can say "Buying on margin in the twenties caused the Great Depression of the thirties", so we can relate segments of discourse causally. We describe the contribution of local coherence information to segmentation, or the recovery of discourse structure, below.

2.2 Discourse Representation Theory

Discourse Representation Theory (Kamp 1981, Kamp and Reyle 1993, Asher 1993, Asher and Kamp 1995) aims to account for the compositional semantic interpretation of entire discourses and to unify its account of truth conditions for discourses with an account of the mental representation of the discourse built by the speaker/hearer.   The discourse representation structure contains separate symbols for each of the distinct events and other abstract types introduced into the discourse. In our approach discourse structure operates over sets of these abstract types.   These sets (discourse segments) form internally coherent eventuality descriptions or propositional content.   Coherence and segmentation inferences operate over these abstract types to produce mental models of coherent discourse. The relata are individual event reference markers, or sums of events (cf. Section 4.1). Coherence predicates are added to the formal discourse representation structure of a text, to form a cognitive model. This is done bottom-up: as each new sentence is interpreted, coherence and segmentation inferences are attempted, and the discourse structure is recovered and modified sentence by sentence.

Our local coherence relation assignment algorithm assigns a coherence relation over two events (or other abstract types) in a cognitive model (Dahlgren, 1989). The algorithm was developed by examining signals of coherence relatedness in a corpus. The information used by the local algorithm includes syntactic properties of the source clause, connectives in either the source or target, the temporal order of the events in the source and target, naive semantic information (world knowledge) associated with the verbs, semantic information such as types of adverbials, mood, and agentiveness. The algorithm was hand tested in the original corpus with 97.5% accuracy.

2.3 Naive Semantics

In order to draw implicit discourses inferences in the absence of overt markers, people employ world knowledge. We briefly describe our lexical theory which incorporates commonsense knowledge to substantiate the feasibility of our approach to coherence. Naive Semantics (NS) (Dahlgren and McDowell, 1986, Dahlgren 1988a, 1991, 1994) is a theory of word sense meaning representation in which associated with each sense of each content word (noun, verb, adjective) is a naive theory of the sort of object, action or property named by the word. The propositions in word sense meanings (e.g., "Cats have tails", "The goal of investing is making a profit") contribute non-monotonically to the truth conditions of sentences of which they become a part. Though not "true", they are and must be close enough to true, enough of the time, for people to refer correctly to real objects and events, and to communicate using language. In principle, the number of feature values could be as large as the number of words in English (Schubert et al. 1979) and thus they do not form a set of primitive concepts from which all others are derived.   This open-endedness is countervailed by the shallowness of lexical knowledge (Dahlgren, 1991).   Naive semantic representations consist only of that shallow knowledge which people must deploy in order to parse and interpret the text, and to find the antecedents of anaphoric expressions.   While the amount of world knowledge associated with a word sense varies with the expertise of the individual, we have found that only a few propositions are required in order to interpret and disambiguate text in many domains, including finance, popular medicine, law, intelligence and software.   We have found that the same shallow lexical knowledge which informs structural and word sense disambiguation is also sufficient for many other forms of reasoning: anaphora resolution (Wada, 1994), coherence (Dahlgren 1989) and discourse segmentation, and relevance reasoning (Dahlgren, 1994).   With independently-motivated naive semantic representations, there is sufficient world knowledge to assign coherence relations for all clauses in Study 1's corpus.

In particular, for verbs, Naive Semantics incorporates the finding that people conceive of actions in terms of their causal implications (Graesser and Clark 1985b).   Naive Semantic representations of verb senses consist of implications such as causes, goals, enablements, consequences, constituent events and states, and typical follow-on events and states, as well as selectional restrictions.   Thus, the psycholinguistically-motivated features in verb representations form the basis of generating coherence inferences.   This is no accident, because they are derived from studies of human text understanding.   For the verb "invest", the goal is to make a profit, the enablement is having money, and so on.   Thus the coherence relation between the investing and profiting events in (1) is derived by inspecting the naive semantic lexicon.

2.4   Other Approaches to Coherence

In comparison with other work, our approach is similar to that of van Dijk and Kintsch (1983).   They define coherence by whether sentences in a discourse describe related facts in some possible world, and they assume that large amounts of world knowledge are employed in building a cognitive model of a discourse. We differ from them in defining coherence as relating discourse events, rather than as relating sentences.   Furthermore, we clarify the question of truth conditions as opposed to naive (or heuristic) inference regarding discourse interpretation.   And we provide an algorithm.   We draw upon Hobbs (1979) and Mann and Thompson (1987) for coherence relations.   However, we define them as relating discourse events (and other abstract types) and structures of these in a cognitive event model, rather than as relating utterances, clauses, or spans of discourse.   The cognitive representation of an event model, after parsing and formal semantic interpretation is complete, is made to cohere and have structure.   For them, coherence is essentially a property of presentation style or of the speaker's intended effect on the hearer.   We agree with Polanyi (1988) that this aspect of coherence belongs in a level of theory above that of cognitive models of interpretation of discourse. For us, coherence is essentially a property of mental models (Johnson-Laird, 1983) which finds its origin in beliefs about relationships among real events. Our approach appeals to cognitive strategies and beliefs people use all the time, whether thinking verbally or not. In the section "Global Coherence and Segmentation", we carefully scrutinize the details of several of these approaches in relation to the question of tree structure for discourse.

3. Approaches To Discourse Segmentation

In contrast to many linguists and sociologists (cf. Chapters 2, Martin, and 1, Schegloff), computational researchers tend to the view that discourse structure is a dominance hierarchy, one that is generative (cf. Hobbs, Chapter 6 of this volume). However, beyond these assumptions there is little agreement about the elements in the dominance hierarchy, the meaning of arcs in the hierarchy, and the level of grammar at which discourse structure is built. In this section we examine the main proposals in discourse theory as they address these foundational issues. A new theory will be proposed which builds discourse structure bottom-up, accounts for anaphora resolution constraints, and gives a unified account of dominance in the discourse tree as topic-relatedness. The theory assumes Asher's (1993) account of event summation and abstract type anaphora. The elements in the theory will be justified in comparison with other proposals.

3.1 Hobbs

Hobbs (1979) builds discourse structure into an undirected graph according to coherence relations between clauses. Higher nodes in the structure are binary coherence relations, and the leaves are clauses. The leaves preserve the sequence of the clauses in the text. An example is the text in (6).

> (6) (a) I had already, as I told you, learned a little bit about hitchhiking,
> (b) I'd split out and two or three times, then come back
> (c) The one—my first trip had been to Geneva New York,
> (d) and then I'd once or twice gone to—twice I'd gone to California,
> (e) And then I'd cut down through the South,
> (f) And I had sort of covered the United States.
> (g) One very beautiful summer... that I spent in Idaho

Hobbs analyzes this monologue clause by clause, finding a coherence relation between each clause and some other clause. Such a procedure accords with Hobbs' convincing claim that if no relation between a new clause and what has come before can be found, the text seems incoherent. The result is tree (7).

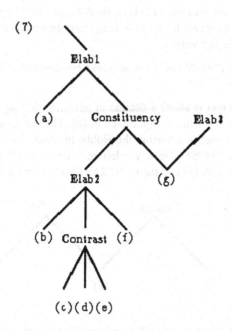

The story in (6) has two main segments, an elaboration of the speaker's hitchhiking experience in clauses (b–f), and an example of one particular trip starting with (g). These segments are readily identified as the subgraphs under ELABORATION 2 and ELABORATION 3 (not shown). Thus the graph captures our intuitive feeling for the structure of the discourse to some extent. However, while the graph preserves linear order and segments are visible, there is no root to the graph, and no semantics of dominance. But in many genres, including some narratives, there is an intended outline to the discourse dominated by a single global discourse topic (Van Dijk and Kintsch 1983, Hinds 1984, Reichman-Adar 1984). This is true for (6), which is clearly about hitchhiking. Thus (a) should be the root of the graph and dominate the structure of the entire narrative. Further, the clause (g) is both an elaboration of clause (a), that is, it describes one hitchhiking event, and is also elaborated by additional clauses (not shown), but it's role relative to both (a) and the additional claues is obscured. In the graph, the node Elab 3 is higher than (g), but (g) is the topic of the remaining (h–m), and should dominate them (Grosz and Sidner 1986, Polanyi 1988).

Another small text, (8), extracted from our *Wall Street Journal* corpus, illustrates the problem with the dominance relation in Hobbs-type discourse structures.

(8) (a) Last fall one of the most promising new shows on television was
       canceled because of the writer's strike.

   (b) In general network viewing is on the decline.

(c) Viewers are reacting to delays, shoddily produced episodes, and changes in schedules by switching from the networks to cable television and videos.

(d) This will probably cost the networks hundreds of millions in revenues.

Intuitively, this text is about a decline in network viewing, exemplified in (a), elaborated in (c), and given an importance in (d). This structure is not transparent in a discourse tree constructed using the Hobbs method. First clauses (a) and (b) are connected via a CONSTITUENCY relation. Then (b) and (c) are connected by ELABORATION (as with (g) and the CONTRAST node dominating (h–m) in (9)).

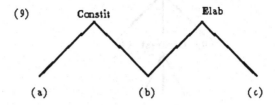

Now where should EVAL clause (d) be inserted? As in the Hobbs tree of (6), we could cross over, yielding the tree in (10). But this tree is most unsatisfying because our intuitive feeling for the structure of the text (and the way it would be segmented by many people) is the tree in (11).

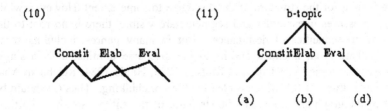

3.2 Our Initial Proposal

In order for the tree to reflect dominance in the discourse structure, so that the higher the node in the tree, the more closely related it is to the global topic (as in VD&K), each node has to be a clause. Then the root can be the topic, nodes one down from the root the main sub-topics in the discourse, and so on.

Clausal Definition of a Discourse Segment

A discourse segment is a possibly discontiguous set of clauses which share a single relation to some other segment or clause. Usually, but not always, segments are contiguous in the discourse surface structure.

We add a new node for each clause and label it with a coherence predicate which relates it to one above. Nodes are clauses and arcs connect them to the clauses with which they most directly cohere in the discourse. Thus, (8) would be represented as in (11). The root of the discourse structure tree becomes the topic of the discourse. Nodes it dominates are those clauses which most directly cohere with it. Those nodes dominate subtrees which elaborate the major sub-topics in the discourse. Higher nodes are more related to the global topic (more salient) in the discourse, subtrees are subsegments, and nodes relate a discourse event or set of events (or other abstract types) mentioned in a dominating node with some event reference marker or set of reference markers introduced at that node. Coherence relations tend to be binary, and this fact is captured by connecting related clauses. In some cases a whole segment relates to a clause.

In some cases, a summary clause introduces or closes the segment. In those cases, the discourse structure is easy to represent. The summary clause can be placed at the node dominating the segment it summarizes. For example, suppose that (8a) is elaborated as in (12).

(12) (a) Last fall "Tattingers" was one of the most promising shows on tv.

(a1) "Tattingers" had one of TV's best producers.

(a2) The star came from another successful series.

(a3) The writers were Emmy winners.

Here the top node of the tree is the generalization (a), and the three elaborations can be attached below it. However, sometimes such subsegments have no introductory or concluding summary. A discourse could start out with examples, and then go on to the topic sentence as in (13).

(13) (a1) "Tattingers" was cancelled.

(a2) "The Cosby Show" was postponed.

(a3) Several new series never materialized.

(b) So it goes in this dismal network season.

In this case, following Polanyi (1988) and Asher (1993), we introduce an implicit topic as in (14).

(14)   (b)

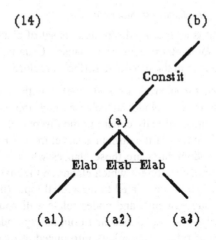

Returning to Hobbs' hitchhiking text in (8), our proposed tree is given in (15).

In (15), the fact that the text is about hitchhiking, and the remainder of the text elaborates this topic, is reflected in the structure of the tree. All nodes are dominated by the topic node (a). Similarly, the elaborations of the generalizations in (b) and (f) are dominated by the nodes which introduce (b) and (f). An arc label is a coherence predicate relating the node below to that above. The nodes are (the content of) clauses. So a coherence predicate elaborate (b,a) can be read off the root. A summary should be higher in the tree, because it is more important and memorable than the details (Van Dijk and Kintsch 1983, Grosz and Sidner 1986).

In the proposed method not only the right-hand side of the tree, but the bottom is open for adding new information to the tree. The analysis of clause (f) would indicate that interior nodes of the tree are open for additions. At the point in the processing when (f) is encountered, a whole subtree has already been built under ELABORATION(b,a), (the subtree covering (c,d,e)). This node is modified by inserting (f) to dominate the subtree (c,d,e), and ELABORATION(f,b). The advantage of permitting additions to the interior of the tree is that the final result will look like an outline of the content of discourse. This makes an empirical prediction that the fewer such corrections and the less backtracking required for its interpretation, the more easily the discourse is processed. It also accords with the finding that in memory for text people build an organization reflecting content (Van Dijk and Kintsch 1983, Morrow et al. 1987). The disadvantage is the computational complexity of redrawing the tree as a restructuring clause is encountered. A less complex mode would keep nodes on the right open, as in Polanyi (1988), and account for reorganization and restructuring of mental models as a later stage of processing.

(15)

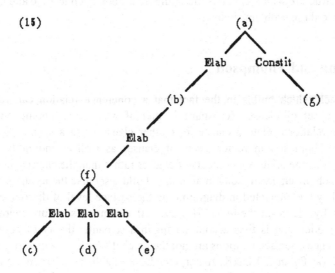

3.3 Polanyi

In Polanyi, the basic discourse unit is the propositional content of a clause, and larger units are built up from these basic units. The nodes are labeled with the propositional content introduced at that node, and lower nodes expand upon the semantic content of higher nodes. If a text implies a generalization that goes unsaid, it is inferred and added to the representation of the discourse. Polanyi assumes that there are only two ways two clauses can be related, as coordinated or subordinated, so that discourse structure trees are binary. As the discourse tree is built up, a new coordinate or subordinate node is added relating a new clause to just one other. In some cases propositions are repeated, and in others they are added to express the content of coordinated propositions. The deep discourse trees it generates do not reflect the intuitive chunks or segments of text which elaborate a topic.

In our theory discourses are broken into segments which are related by coherence predicates. The discourse structure is built up from the expressed content, topic inferences are only added if required by the discourse structure. In contrast, Polanyi's binary parsing method forces topic inferences for every pair of clauses under an subordination node where no such topic is directly expressed. Another problem with Polanyi's theory is that the root of the binary tree does not in general correspond to the global discourse topic, and subtrees to subtopic. Height in the tree is unrelated to topic. In contrast, in our theory the tree structure reflects outlining. The final problem with the Polanyi theory, which holds as well for Hobbs, is that it places clauses at the nodes. This leads to an overly fine-grained discourse structure, and collapses the distinction between syntactically signaled semantic properties (embedding, conditionals, quantifiers) and the discourse structure which explains the intuitive notions of coherence and outlining. We return to these problems below.

3.4 Mann and Thompson

An approach which builds in the fact that a coherence relation can relate one clause to a set of clauses (or span) is that of Mann and Thompson (1987). Coherence relations relate a clause to another clause or to a span.. We follow Mann and Thompson in relating sets of clauses as well as individual clauses, although for us the relata are contents of clauses rather than the clauses themselves. Our approach differs from theirs in aiming to build discourse theory upon semantic theory. They are interested in diagramming the surface level of discourse, clause-in-the-text by clause-in-the-text. However, if discourse structure belongs in a cognitive model, and is thus accounted for in a semantic theory of content plus naive inferences, surface relations are not the level at which discourse structure can be defined. (as Polanyi, VD&K, Asher, etc., have established). Second, dominance in the diagrams is unrelated to importance in the text, and the topic is not necessarily at the top. Third, the spans must be contiguous in the text, and the leaves of the diagram are the clauses of the text in the order they appear. The discussion of the Hobbs text (6) shows where the discourse structure doesn't always follow exactly the order of clause presentation.

3.5 Grosz and Sidner

Our discourse segmentation method is similar to that of Grosz and Sidner (1986), hereafter G&S. The segmentations result in a hierarchy in which the root is the discourse topic, and subtrees correspond to subtopics. Furthermore, the trees are non-binary. Our treatment of anaphor resolution differs in some respects, but agrees with G&S that discourse structure is an important determinant of the availability of discourse entities as antecedents of anaphors. There are two key differences. First, G&S are concerned with explaining the effect of speaker intentions on the discourse structure, while we are concerned with the recovery of the semantic content of the discourse, and assign intentions and planning to a higher-level theory. This follows Polanyi (1988), where the levels are: discourse structure, genre, speech event and interaction. Speaker intention belongs at the level of speech event, and is one of the determinants of genre. Secondly, G&S abjure coherence relations as under-defined (a problem this paper hopes to address). Our Study 1 of discourse segmentation (described below) shows that coherence relations are fundamental to the recovery of discourse structure, at least in some genres. In newspaper commentary, change in coherence relation is the main signal of change of discourse segment in a commentary genre, while explicit cue phrases, are infrequent. In a corpus of 16000 words, an average of 90% of segment boundaries change discourse relation, while only 16% have a cue phrase. Coherence is the most informative factor determining discourse segmentation and thus constraining anaphora resolution. Even though the theory of coherence relations is at a formative stage, it is a necessary element of any theory that hopes to account for the facts.

In G&S, the two example texts illustrate the fact that genre is an important element of the theory of discourse. Their examples belong in the rhetoric and task-oriented genres. In the commentary genre we have studied, the topic-dominated discourse structure tree is explained by and determined by coherence reasoning about events. In other genres, the structure is very different. In the rhetoric text, the topic is often a proposition, as in G&S's example, where the topic proposition is "Parents shouldn't allow children to go to movies". The segments dominated by the root are background and arguments justifying the topic proposition in some logical argument form such as syllogism, induction, etc. On the other hand, the pump example is in the task-oriented dialogue genre, where the discourse structure is determined by the structure of a task (and even, to some extent, by the structure of the artifact being built), rather than by speaker hypotheses concerning the coherence of events.

3.6 Our Revised Proposal

Thus, the revised version of discourse structure theory we propose is as follows:

> The propositional content of clauses is placed at nodes. Each node is labeled with a coherence relation involving the content at that node, and a higher node. Thus arcs in the tree signify that the nodes they connect are coherence related. Height in the tree corresponds to topic-relatedness in some genres. The deeper the tree, the more complex the outline, and the longer the discourse. Subtrees are segments of the discourse. The global topic of the discourse is at the root of the tree. Interruptions and repairs are not part of the discourse structure tree.

Below, we modify this theory to make trees shallower by   recognizing event summation in cognitive representations.

3.7 Study 1 of Naive Semantics and Discourse Segmentation

The purpose of the study was to discover the signals of discourse segment boundaries. The study was empirical in that a corpus of Wall Street Journal commentary was analyzed and segmented, the potential segmenting factors were assigned for all clauses (those at segment boundaries and those not), and the information content level of each factor was tabulated. Coherence relations and segmentation were intuited. A check on segmentation boundaries was made by having two judges segment independently. However, as pointed out in Chapter 7 (Passonneau and Litman), considerable variability exists in segmentation.

A discourse segment for purposes of this study is a chunk of a discourse which has the same outermost coherence relation to some other discourse entity or segment. In other words, clauses deeply embedded inside a subtopic share a coherence relation to the topic with the dominating subtopic node and all of its daughters. They have a unified function in relation to the rest of the discourse. Two judges (the author and Carol Lord) segmented 16000 words of WSJ commentary using the method of G&S (non-binary trees, topic at the root). The two judges agreed on all but one of the segment boundaries. In what follows, "new sister" refers to adding a new sister node to the discourse structure tree as in adding node 3 to (16a), resulting in (16b).

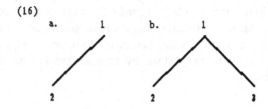

(16)

"New subsegment" refers to adding node 4 to the tree in (16a), resulting in the tree of (17a). In terms of text, a new sister looks like the bottom portion of (17b), and a new subsegment like the innermost portion of (17c).

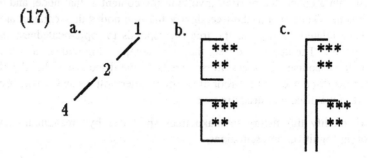

(17)

The author then examined factors which might cause the reader to segment the discourse at new segment boundaries, as well as factors which are present when the reader does not change segment. The cues which seem likely to be important in signalling a new discourse segment are new paragraph indentation, discourse cues such as "Turning to...", "In summary", the use of event anaphors and changes in coherence relation, sentence subject and tense. We wanted to see whether pronoun use could be predicted by segmentation algorithm which ignored pronoun use in locating segment boundaries. Table 1 shows the results.

Table 1. Signals of New Segments in a 16000 word corpus

	New Sister Segment	New Subsegment	Non-segmenting
Change in coherence relation	92%	88%	<1%
Change of sentence subject	89%	70%	52%
Segmenting cue phrase	16%	11%	—
Event anaphora	29%	9%	13%
Change in tense or aspect	49%	53%	30%

The most consistent correlate of new segment was change in coherence relation. Change in coherence relation is a stronger predictor than change in sentence subject, because while the latter is present in 89% of new sister segments and 70% of new subsegments, it is also present in half of the clauses which do not change segment. In contrast, change of coherence relation is almost never present when there is no change in segment. It was surprising that cue phrases are relatively unimportant, though clear, indicators of change of segment. Segmenting cue phrases were present in only 16% of the clauses which changed segment. On the other hand, event anaphors proved to be more important. Event anaphors are pronouns such as "this" or "it", or definite NP's such as "the move", which have events as antecedents. Frequently the antecedents are segments of the discourse, and the event anaphor makes the reader segment the previous discourse correspondingly. (This fact supports Asher's (1993) argument that event anaphoric phenomena are indicative of the processes involved in discourse segmentation.) Change in tense or aspect is not an informative factor in segmentation, except insofar as it contributes to coherence relation assignment.

Anaphoric devices (both eventive and non-eventive) are also important cues in segmentation. Our Study 2 reveals several important distinctions turning on type of anaphoric device and type of referent. Individual type referents of anaphors must be inside the segment, but event type referents tend to be outside it. Furthermore, definite NP anaphors can refer anywhere in the text, but the further away the antecedent the "heavier" the anaphoric NP. The study also confirms G&S's point that pronouns do not refer outside the focus stack (that is, they refer either inside the segment or to an entity introduced in a dominating clause in the discourse). In fact, we found that all individual pronouns in the WSJ corpus had referents in the immediately preceding sentence inside the segment. We studied the positions of anaphoric devices and their antecedents in the 16000 words of WSJ corpus which we segmented. Our findings were:

- Pronouns had antecedents in the immediately preceding sentence in the same segment.
- Demonstratives with event antecedents had antecedents outside the segment, but demonstratives with individual antecedents did not.
- Definite NP's have antecedents in other segments.
- There is a correlation between the heaviness of an anaphor and the distance (in number of sentences) back to its antecedent.

The heaviness hierarchy is:

- pronoun
- Det NP
- Det (Adj) NP (PP)
- Det (Adj) NP (Participial)
- Det (Adj) NP (Relative Clause)

First, these findings corroborate Asher's (1993) fundamental distinction between event and individual anaphora. Here the constraints on event and individual anaphora are very different. Second, these findings support a method of segmentation to be described below, because they call for constraints on anaphors which can be simply stated with relation to the proposed tree structure. The constraints are:

Constraint on Individual Anaphora

The antecedent must be in local sister leaf node (minimal governing node which c-commands the node with the pronoun), or in the global topic.

Constraint on Abstract Type Anaphora

The antecedent must be in a sister, not necessarily local, or some dominating node.

Constraint on Heavy Definite NP Anaphora

Unconstrained.

Thus, abstract type anaphors may find their antecedents up the tree at any point, while individual anaphors must find their antecedents locally or at the root. These constraints are modified below in light of the revised discourse structure theory. We also found that the notion of topic was substantiated by the facts of anaphor resolution in the WSJ corpus. Lord counted the distance back to the antecedent for all anaphors (pronouns, demonstratives, and definite NPs), using 23 topic event, 13 non-topic event, 84 topic event participant, and 80 non-topic event participants. The generalization held that antecedents in topic events and topic event participants could be located farther away than other antecedents. Looking at event anaphors, for non-topics, the antecedent was typically in the previous sentence and in the same segment. For topics, the antecedent was sometimes in the previous sentence, but more often back two or more sentences, and in a previous segment. Looking at individual anaphors, the antecedent was typically in the previous sentence and the same segment, but the likelihood of its being farther back was greater for topic event participants than for non-topic event participants.

4. Improving Our Proposal

The trees we have presented so far do not tell the whole story of discourse structure. We would like to have the trees reflect the anaphora resolution constraints. In our corpus, all individual anaphors had antecedents in the previous sentence, and all but two of the antecedents were inside the same segment. However, we know that in general anaphoric pronouns are not as constrained in their use as that. It is easy to imagine a segment in which an individual is introduced, not a topic participant, followed by a complex rhetorical structure, such as an argument, followed by a sentence with a personal pronoun. G&S introduce focus spaces, and Reichman (1985) context spaces, on the theory that topic-related (or plan-related) segments constrain anaphor resolution. In their examples, there is no case of a pronoun in a sister or closed focus space. We see no reason, in expository text (as opposed to task-oriented dialogue), that such examples should not be found, as long as a certain general distance and same-gender constraints are upheld. In task-oriented dialogue, it makes sense that certain objects which are no longer relevant because they have been set down on the work bench, as in the pump example (G&S), should not be available. However, in expository text, or narrative, it is possible to introduce a person or thing, continue on a related or background subject or argument, and then refer to the introduced person or thing with a pronoun. Consider the following text:

(18) (a) John likes to invent computer games.
(b) He often collaborates with his sister to come up with unusual logics.
(c) He started out with games like Dig Dug and Mario.
(d) These games fascinated him,
(e) but after a short time he would pick up the logical structure, and
(f) become bored.
(g) She felt the same way...

In this example, John is a topic participant, but his sister isn't. Using the trees we've been proposing so far, (18) is analyzed as in (19).

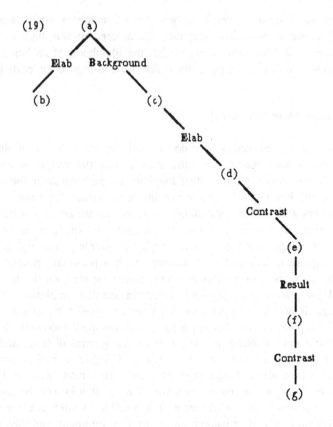

The antecedent of "She" in clause (g) is found in clause (b). Thus (18) is a counterexample to our constraint on individual pronominal anaphora resolution. In G&S terms, the segment labeled (b) in (19) should have popped from the focus stack because it does not dominate the segment starting at (c). Though (18) is a counterexample to the constraint, some other constraints must replace it. We know that the antecedent has to be close enough for human memory processing to be able to recover the antecedent.   Further, if same-gender, commonsensically plausible alternatives intervene, it becomes hard for the hearer to recover the antecedent. On the other hand, it is clear that a rule which examines finely grained local-coherence based discourse trees cannot capture the empirical facts concerning pronominal anaphora resolution constraints, which allow for looking back into "popped" local segments.

An independent reason to reject clause-based or propositional content of clause-based discourse structure theory is that it ignores the distinction between semantics and discourse structure.   Embedding phenomena, including belief contexts, conditionals and others, universal quantifiers, and any other phenomena with more than one clause in a sentence, must be taken into account by the discourse structure representation in a clause-based theory.

In considering how to alter our theory, we must remember that text segmentation varies with genre, and the example from task oriented dialogue discussed by G&S is unusual in having extremely localized focus spaces. In other genres, such as commentary, expository pedagogy (textbooks), and narrative fiction, segmentation operates to chunk text into topic-related segments. The typical novel segments into chapters. Chapters can have (though they need not have) topics dealing with background, action sequences, persons and so on. Within the chapter there are smaller segments, some of which may have topics. These may also have subsegments, forming a dominance hierarchy. The point is that unlike task-oriented dialogue, inside certain-sized chucks, anaphoric pronouns don't function according to the focus space constraints suggested by G&S, as exemplified in (18).

Our discourse trees must be relatively shallow, then, in order to account for availability. Both length and genre must be considered in segmentation. We need a theory which says that in the commentary genre, segments are topic-related, dominance in the discourse tree is topic dominance, and leaves of the discourse structure tree are chunks of text of a certain length (determined by memory processing constraints, and correlated with the possibility of recovering antecedents of individual pronominal anaphora). For a commentary text like that summarized in Section 4.4, this means a discourse structure like (20).

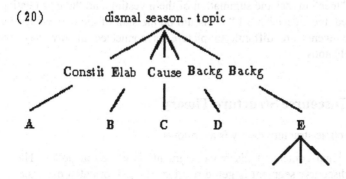

The segments labeled A–D are summed events (states, propositions, etc.) and the length in the surface structure is short enough that personal pronouns can find their antecedents within them. In the text of Section 4.4, segment E contains relatively long subsegments dealing with the decline of the three networks. Each subsegment has an elaborative structure in which pronouns can find their antecedents. (Noting, of course, that in this particular text, personal pronouns all had antecedents in the preceding sentence.) Thus, our theory must recognize length in segmentation, as well as a recursive topic related structure. Exactly the length limit for a segments which predict individual pronoun anaphora is an empirical question concerning human memory processing constraints.

4.1 Event Summation

Asher (1993) proposes a semantic treatment of abstract objects including events, states, facts, propositions and concepts. He shows that the pronouns "it", "this" and "that" can have as antecedents chunks of discourse which are summed events (or sums of other abstract objects). An example of such an anaphoric usage is found in the text in Section 4.4, where the demonstrative "so" has as antecedent an event type corresponding to "the kind of thing that happened to the Tattingers show". Such summing event anaphors abound in our corpus of commentary text. Asher also shows that events, propositions and other abstract types are antecedents of pronouns more locally, and proposes a theory which also accounts for various syntactic phenomena such as verb phrase ellipsis. He demonstrates that event summation is a necessary part of any semantic theory. He then goes on to define event summation semantically within Discourse Representation Theory (Kamp 1981), as the join of discourse events which can plausibly occur in an event chain. Built into the theory is the recognition of naive semantic constraints on event summation. A felicitous discourse will not force the hearer to sum events which in naive semantics are unlikely to be related. An example of such an infelicitous discourse is found in (21).

(21) John invested heavily. Then he ate pizza. It was gross.

The cue phrase "then" invites the summation of the investing and the pizza eating, and if so summed, the antecedent of "it" should be the sum of the two events, but since these two events are difficult to picture as connected in any way, the discourse is infelicitous.

4.2 Revised Discourse Structure Theory

Our revised discourse structure theory is as follows:

The propositional content of discourse segments is placed at nodes. The length of a discourse segment is genre-relative. In task-oriented dialogue, discourse segments are any change of coherence relation, while in commentary and narrative, length is a size chunk in which the availability of antecedents for personal pronouns holds. Each node is labeled with a coherence relation involving the content at that node, and a higher node. Thus arcs in the tree signify that the nodes they connect are coherence related. Height in the tree corresponds to topic-relatedness in some genres. The deeper the tree, the more complex the outline, and the longer the discourse. Subtrees in topic genres are subtopics. The global topic of the whole discourse is at the root of the tree. Internal to the discourse segment in certain genres is recursively-built coherence structure among abstract types which does not constrain anaphora resolution. Interruptions and repairs are not part of the discourse structure tree.

This theory differs from the clausal version. The changes incorporate event summation, and shallowness of discourse structure to account for anaphora resolution constraints. The resulting anaphor resolution constraints reflecting the revisions in definition of the trees are given below.

Constraint on Individual Anaphora

The antecedent must be inside the same segment, or in the global topic.

Constraint on Event (Abstract Type) Anaphora

The antecedent must be in a local sister or in a dominating node.

4.3 Formal Discourse Structure

Each clause introduces individual, event, state, and propositional reference markers into the discourse representation structure (DRS). (22) is the DRS representing the text "John invested heavily. He made a huge profit". Temporal relations between events are reflected in the DRS, and anaphoric relations are determined. Coherence can be viewed as defined over constituents of the DRS which reflect the content introduced by clauses.

(22)

$u1,e1,u2,u3,e2,r1,r2$
$r1 <$ now
john(u1)
e1 invest(u1)
e1 incl in r1
heavily(e1)
profit(u3)
e2 make(u2,u3)
huge(u3)
$r1 < r2$
e2 included in r2
u2=u1

The constituent K1 corresponding to John's investing is as in (23). A Discourse Structure (DS) represents the coherence of a DRS.

(23)

$u1,e1,r1$
$r1 <$ now
john(u1)
e1 invest(u1)
e1 incl in r1
heavily(e1)

In (24) we see the DS for (23). Here the two constituent DRS's introduced by John's investing (K1) and John's making a profit (K2) are causally related.

(24)

topic(K1,goal(K1,K2))

In order to integrate a DRT approach with the principles stated in Section 4.2, we must again revise those principles:

Definition of Discourse Segment

A node of the discourse structure is a portion of a DRS, that is, some (possibly non-contiguous in surface structure) subset of the reference markers and corresponding conditions in the DRS.

The content of that constituent DRS plays a topic role in the discourse. The constituent DRS comes from a portion of the surface discourse which does not exceed maximum memory capacity for finding antecedents of personal pronouns.

Arcs are labeled with binary coherence relations.

The following principle must be added:

Discourse segment boundaries occur where there is a change in coherence relation at the segment level.

That is, where it is no longer possible to continue summing the events, states and other abstract reference markers in such a way that the sum has the same coherence relation as before to the remainder of the discourse (in particular, to the node above in the discourse structure tree), there is a discourse segment boundary. Reflecting these changes, the anaphora resolution constraints become:

Constraint on Individual Pronominal Anaphora:

Antecedent reference markers must be in an accessible position inside the same segment, or in the global topic.

Constraint on Abstract Type Pronominal Anaphora

Antecedents may be any abstract type reference marker (event, state, fact, proposition), concept type reference markers (delineated DRS's), DRS's or discourse segments found in a local sister, or in a dominating node in the discourse structure tree.

This theory predicts that change of discourse relation results in change of segment, one of the main findings in our Study 1.

The approach proposed here predicts the availability of antecedents for anaphors by both the formal semantic constraints of Discourse Representation Theory, and the constraints of the discourse segment structure. Segments are signalled by coherence relations and coherence is defined in terms of naive theories of event relatedness. The definition of local coherence is essentially psychological (cf. Graesser and Clark 1985b).

Global coherence is defined by essentially the same intuitions. People do segment discourse, especially certain genres, they do infer causal relations between the events reported in chunks of discourse. Furthermore, written expository text is planned and outlined by writers with segments in mind (and often explicitly marked with subtitles).

The coherence algorithm draws upon local linguistic cues plus the naive semantic lexicon. The segmentation algorithm draws upon local coherence information. Change of segment hypotheses are generated when one of the significant factors is encountered. These factors, in order of importance are: change of local coherence, change of subject, paragraph indentation, segmenting event anaphoric device segmenting cue phrase.

4.4 An Example

An example analysis of a text according to the final theory will be given here regarding the text sketched in the following table. (Space does not permit reproduction of the entire text). The topic-related segments are identified by discourse variables K1–K10. These variables correspond to subsets of the reference markers and conditions in the root DRS for the entire text. K1 would identify the reference markers and conditions for the content of sentences 1–5, and so on as listed below. K11 corresponds to the content of the topic sentence, sentence 6.

Vars	Sents	Text topics
K1	1–5	The promising Tattinger's show was canceled.
K2	6–10	So it goes in this dismal network season.
K3	11–14	This is taking place while television watching is on the rise.
K4	15–22	Fall debuts were delayed by the writer's strike.
K5	23–30	The studios may save millions from concessions by the strikers.
K6	31–34	This is the latest in a string of declines in network ratings.
K7	32...	CBS is hurting the worst.
K8	33...	No. 1 NBC has seen its audience shrink 7% overall.
K9	34...	ABC is the only network doing better.
K10	35–41	The strike's impact will be felt well into the next TV season.
K11	6	So it goes in this dismal network season.

The corresponding discourse structure tree would be as in (25).

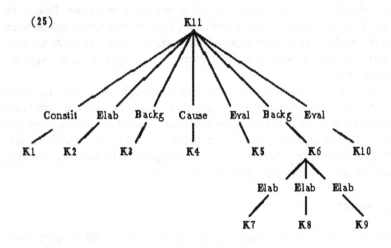

(25)

5. Conclusion

We hope to have demonstrated that coherence can be explained in terms of cognitive strategies which are used to make sense of events and event chains, and structures of other types such as states, mental states, propositions and so on. A plausible architecture is one in which coherence inferences operate upon the logical form of a discourse and employ commonsense knowledge. Local coherence relations can be extracted using syntactic, semantic and naive semantic information. Knowledge representation for naive semantics (Dahlgren 1988a, 1991, 1994) is adequate to drive almost all of the coherence relation assignments, including implicit ones (Dahlgren 1981). Global discourse structure is genre relative (Polanyi 1988) and topic-related (Van Dijk and Kintsch 1983). Constraints on anaphora resolution recognize topic-related discourse segments. Hence global discourse structure trees must be relatively shallow.

Acknowledgments

I want to thank Nicholas Asher, Clair Chyi, Carol Lord, Art Graesser, Joyce P. McDowell, Edward P. Stabler, Jr. and Hajime Wada for discussions of these issues and commentary on this work. Lord (at Intelligent Text Processing, Inc.) and Stabler developed the parser and McDowell and Wada the DRT component of the natural language system (InQuizit) in which the coherence algorithms described here operate.

References

Asher, N. 1993. Reference to Abstract Objects in Discourse. Norwell, MA: Kluwer.

Asher, N. and Kamp, H. 1995. Discourse Structure in the Service of Lexical Disambiguation. Ms in prep.

Cohen, R. 1984. A Computational Theory of the Function of Clue Words in Argument Understanding. *Proceedings of the COLING-84 Conference*, 251–258.

Dahlgren, K. 1985a. The Cognitive Structure of Social Categories. *Cognitive Science* 9, 379–398.

Dahlgren, K. 1988a. *Naive Semantics for Natural Language Understanding*. Boston, MA: Kluwer.

Dahlgren, K. 1988b. Using Commonsense Knowledge to Disambiguate Word Senses. In P. St Dizier and V. Dahl (eds.), *Natural Language Understanding and Logic Programming 2*. Amsterdam: North-Holland.

Dahlgren, K. 1989. Coherence Relations and Naive Semantics. Paper given at the Symposium on Modelling Discourse Structure: Discourse Segments and Discourse Relations, University of Texas, Austin.

Dahlgren, K. 1990. Naive Semantics and Robust Natural Language Processing. *Proceedings of the AAAI Spring Symposium*.

Dahlgren, K. 1991. The Autonomy of Shallow Lexical Knowledge. In J. Pustejovsky (ed), *Knowledge Representation and Lexical Semantics*. New York: Springer-Verlag.

Dahlgren, K. 1992. Convergent Evidence for a Set of Coherence Relations. In D. Stein (ed.), *Cooperating with Written Texts*. New York: Mouton de Gruyter.

Dahlgren, K. 1993. A Linguistic Ontology. *Proceedings of the International Workshop on Formal Ontology*.

Dahlgren, K. 1994. Finding Relevant Texts. *Proceedings of the AAAI Fall Symposium*, New Orleans.

Dahlgren, K. and J. McDowell. 1986a. Kind Types in Knowledge Representation. *Proceedings of the COLING-8 Conference*.

Dahlgren, K., J.P. McDowell, and E.P. Stabler, Jr. 1989. Knowledge Representation for Commonsense Reasoning with Text. Forthcoming in *Computational Linguistics*.

Davidson, D. 1967a. Causal Relations. *Journal of Philosophy* 64, 692–703.

Davidson, D. 1967b. The Logical Form of Action Sentences, in N. Rescher (ed.), *The Logic of Action and Preference*.

Decker, N. 1985. The Use of Syntactic Clues in Discourse Processing. *Proceedings of the ACL-85 Conference*.

Foss, D.J. 1988. Experimental Psycholinguistics. *Annual Review of Psychology* 30, 301–348.

Graesser, A. and L. Clark. 1985a. *Structure and Procedures of Implicit Knowledge*. Norwood, NJ: Ablex.

Graesser, A. and L. Clark. 1985b. The Generation of Knowledge-Based Inferences during Narrative Comprehension. In G. Rickheit and H. Strohner (eds.), *Inferences in Text Processing*. Amsterdam: North-Holland.

Grosz, B. and C. Sidner. 1986. Attention, Intensions and the Structure of Discourse: A Review. *Computational Linguistics* 7, 85–98; 12, 175–204.

Hayes, P. J. 1985. The Second Naive Physics Manifesto. In J.R. Hobbs and R.C. Moore (eds.), *Formal Theories of the Commonsense World*. Norwood, NJ: Ablex.

Hinds, J. 1977. Paragraph Structure and Pronominalization. *Papers in Linguistics* 10(1–2), 77–99.

Hirst, G. 1981. Discourse-Oriented Anaphora Resolution: A Review. *Computational Linguistics* 7, 85–98.

Hobbs, J.R. 1979. Why is Discourse Coherent? SRI Technical Note #176.

Hobbs, J.R. 1985. On the Coherence and Structure of Discourse. CSLI Report #CSLI-85-37.

Holland D. and N. Quinn, eds. 1987. *Cultural Models in Language and Thought.* Cambridge: Cambridge University Press.

Hovy, E.H. 1990. Parsimonious and Profligate Approaches to the Question of Discourse Structure Relations. *Proceedings of the 5th International Workshop on Text Generation.*

Johnson-Laird, P.N. 1983. *Mental Models.* Cambridge: Harvard University Press.

Kamp, H. 1981. A Theory of Truth and Semantic Representation. In J. Groenendijk, Th. Janssen, and M. Stokhof (eds.), *Formal Methods in the Study of Language.* Amsterdam: Mathematisch Centrum, 277–322.

Lakoff, G. 1985. *Women, Fire and Dangerous Things.* Chicago, IL: University of Chicago Press.

Mann, W.C. and S.A. Thompson. 1987. Rhetorical Structure Theory: A Theory of Text Organization. ISI Reprint Series: ISI-RS-87-190.

Morrow, D.G, S.L. Greenspan, and G.H. Bower. 1987. Accessibility and Situation Models in Narrative Comprehension. *Journal of Memory and Language* 26, 165–187.

Partee, B. 1984. Nominal and Temporal Anaphora. *Linguistics and Philosophy* 7, 243–286.

Polanyi, L. 1988. A Formal Model of the Structure of Discourse. *Journal of Pragmatics* 12, 601–638.

Reichman-Adar, R. 1984 Technical Discourse: The Present Progressive Tense, The Deictic 'That' and Pronominalization. *Discourse Processes* 7, 337–369.

Reiter, R. 1980. A Logic for Default Reasoning. *Artificial Intelligence* 13, 81–132.

Rosch, E., C.B. Mervis, W.D. Gray, D.M. Johnson, and P. Boyes-Braem. 1976. Basic Objects in Natural Categories. *Cognitive Psychology* 8, 382–439.

Schubert, L. K., R.G. Goebel, and N.J. Cercone. 1979. The Structure and Organization of a Semantic Net for Comprehension and Inference. In N.V. Findler (ed.), *Associative Networks.* New York: Academic Press.

Smith, E.E. and D.L. Medin. 1981. *Categories and Concepts.* Cambridge, MA: Harvard University Press.

Strawson, P.C. 1953. *Individuals.* London: Methuen.

Trabasso, T. and L.L. Sperry. 1985. Causal relatedness and the Importance of Story Events. *Journal of Memory and Language* 24, 595–611.

Trabasso, T. , T. Secco, and P. van den Broek. 1983. Causal cohesion and story coherence. In H. Mandl, N.S. Stein and T. Trabasso (eds.), *Learning and Comprehension of Text.* Hillsdale, NJ: Erlbaum.

Van Dijk, T. and W. Kintsch. 1983. *Strategies of Discourse Comprehension.* New York: Academic Press.

Vendler, Z. 1967. *Linguistics in Philosophy.* Ithaca: Cornell University Press.

Wada, H. 1994. A Treatment of Functional Definite Descriptions. *Proceedings of the Coling-94 Conference.*

Wilks, Y. 1975. Preference Semantics. In E. Keenan (ed.), *Formal Semantics of Natural Language.* Cambridge: Cambridge University Press.

Chapter 6
On the Relation Between the Informational and Intentional Perspectives on Discourse

Jerry R. Hobbs

SRI International, Menlo Park

1. Introduction

In the paper "Interpretation as Abduction" (hereafter IA) Hobbs et al. (1993) presented and elaborated the view that to interpret an utterance is to find the best explanation of why it would be true. We may call this the "Informational Perspective" on discourse interpretation. The only thing to be explained is the information explicitly conveyed by the utterance, and the explanation does not necessarily involve any knowledge of the specific goals of the speaker.

Norvig and Wilensky (1990) raised the objection to this approach that what really needs to be explained is what the speaker was trying to accomplish with the utterance. Under this view, to interpret an utterance is to find the best explanation of why it was said. We may call this the "Intentional Perspective" on discourse interpretation.

The Intentional Perspective has been the canonical view in natural language processing since the middle 1970s. It originated with Power (1974), Bruce (1975), and Schmidt et al. (1978), and is the view adopted in Cohen and Perrault (1979), Grosz (1979), Allen and Perrault (1980), Perrault and Allen (1980), Hobbs and Evans (1980), Grosz and Sidner (1986) and many others since that time. The view taken in all of this work is that the speaker is executing a plan, the utterance is an action in that plan, and the job of the hearer is to discover the plan and the role that the utterance plays in the plan. This is an especially useful, indeed essential, perspective when the discourse is a dialogue in which most turns are a sentence or less in length and the participants' plans are being modified continuously by the interaction.

It is clear why the Intentional Perspective is the correct one when we look at things from the broadest possible point of view. An intelligent agent is embedded in the world and must, at each instant, understand the current situation. The agent does so by finding an explanation for what is perceived. Put differently, the agent must explain why the complete set of observables encountered constitutes a coherent situation. Other agents in the environment are viewed as intentional, that is, as planning mechanisms, and this means that the best explanation of their observable actions is most likely to be that they are steps in a coherent plan. Thus, making sense of an environment that includes other agents entails making sense of the other agents' actions in terms of what they are intended to achieve. When those actions are utterances, the utterances must be understood as actions in a plan the agents are trying to effect. That is, the speaker's plan must be recognized—the Intentional Perspective.

But there are several serious problems with the Intentional Perspective. First, the speaker's plan can play at best an indirect role in the interpretation process. The hearer has no direct access to it. It plays a causal role in some observable actions, in particular the utterance, which the hearer can then use, along with background knowledge, to form a belief about exactly what the plan is. Only this belief can play a direct role in interpretation. How is the hearer to arrive at this belief? How can the hearer go from utterance to intention, in those cases where there is no prior knowledge of the intention?

There is a further problem, that occurs especially in extended, one-speaker discourse, such as written text. There is a level of detail that is eventually reached at which the Intentional Perspective tells us little. It tells us that the proper interpretation of a compound nominal like "coin copier" means what the speaker intends it to mean, but it offers us virtually no assistance in determining what it really does mean. Frequently what the speaker intends an utterance to mean is just what it would mean if spoken by almost anyone else in almost any other circumstance. We need a notion of interpretation that is independent of and goes beyond speaker's intention. It must, for example, give us access to plausible relations between coins and copiers.

A third problem with the Intentional Perspective is that there are many situations in which the speaker's plan is of little interest to the hearer. Someone in a group conversation may use a speaker's utterance solely as an excuse for a joke, or as a means of introducing a topic he or she wants to talk about. Very often two speakers in a discussion will try to understand each other's utterances in terms of their own frameworks, rather than attempt to acquire each other's framework. A medical patient, for example, may describe symptoms according to some narrative scheme, while the doctor tries to map the details into a diagnostic framework. A spy learning a crucial technical detail from the offhand remark of a low-level technician doesn't care about the speaker's intention in making the utterance, but only about how the information fits into his own prior global picture. A historian examining a document often adopts a similar stance. In all these cases, the hearer has his or her own set of interests, unrelated to the speaker's plan, and interpretation involves primarily relating the utterance to those interests.

In brief, the role of the speaker's intention is indirect, it is often uninformative, and it is frequently not very important. It cannot be the whole story. We need to have an intention-independent notion of interpretation.

Our first guess might be that we simply need the literal meaning of the utterance. But an utterance does not wear its meaning on its sleeve. Anaphora and ambiguities must be resolved. Metonymies and ellipsis must be expanded. Vague predications, including those conveyed by the mere adjacency of words or larger portions of text, must be made specific. In short, the utterance must be interpreted. The notion of literal meaning gets us nowhere.

A primary use of language is to present the facts about a situation. To understand a situation that we perceive we have to find an explanation for the observable facts in that situation. Similarly, to understand a situation that is described to us we must find an explanation for the facts we are told. But this is exactly the account of what an interpretation of an utterance is under the Informational Perspective. The "informational interpretation" gives us an analogue of literal meaning that is adequate to the task. As shown in IA, interpreting an utterance by finding the best explanation for the information it conveys solves as a by-product the problems listed above—resolving anaphora and ambiguities, expanding metonymies and ellipsis, and determining specific meanings for vague predicates.

The informational interpretation is, to be sure, relative to an assumed background knowledge. Conversation is possible only between people who share some background knowledge, and interpretation is always with respect to some background knowledge that the hearer presumes to be shared. The explanation that constitutes the interpretaton has to come from somewhere. But conversation, and hence interpretation, is possible in the absence of information about the other's specific goals. We have conversations with strangers all the time.

The picture that emerges is this. Humans have constructed, in language, a tool that is primarily for conveying information about situations, relying on shared background knowledge. Like all tools, however, it can be put to uses other than its primary one. We can describe situations for purposes other than having the hearer know about them. The Informational Perspective on discourse interpretation tells us how to understand the situations described in a discourse. The Intentional Perspective tells us how to discover the uses to which this information is being put.

The Intentional Perspective on interpretation is certainly correct. To understand what's going on in a given communicative situation, we need to figure out why the speaker is making this particular utterance. But the Informational Perspective is a necessary component of this. We often need to understand what information the utterance would convey independent of the speaker's intentions. Another way to put it is this: We need to figure out why the speaker uttered a sequence of words conveying a particular content. This involves two parts, the informational aspect of figuring out what the particular content is, and the intentional aspect of figuring out why the speaker wished to convey it.

It should not be concluded from all of this that we first compute an informational interpretation and then as a subsequent process compute the speaker's intention. The two intimately influence each other. Sometimes, especially in the case of long written texts and monologues, the informational aspect completely overshadows considerations of intention. Other times, our knowledge of the speaker's intention completely masks out more conventional readings of an utterance. We consequently need a framework that will give us the conventional meaning, relative to a shared knowledge base, but will also allow us to override or to completely ignore this meaning when more is known about the speaker's aims. This paper is a preliminary effort to provide such a framework.

In Section 2 the IA framework is presented, in just enough detail to allow this paper to stand on its own. The interested reader should consult IA and the other cited papers for a deeper discussion of the framework. In Sections 3 and 4, two examples are given. In the first, the informational reading and the intentional reading are essentially the same. In the second, they are in conflict and the intentional reading wins out. Section 5 summarizes.

2. Background

2.1 Logical Notation

In this paper I will use the ontologically extravagant, first-order notation introduced in Hobbs (1983, 1985, 1995). For the purposes of this paper, the chief feature of this notation is the use of eventualities.

We will take $p(x)$ to mean that p is true of x, and $p'(e,x)$ to mean that e is the eventuality or possible situation of p being true of x. This eventuality may or may not exist in the real world. The unprimed and primed predicates are related by the axiom schema

$$(\forall x)\, p(x) \equiv (\exists\, e)\, p'(e,x) \wedge Rexists(e)^1$$

where $Rexists(e)$ says that the eventuality e does in fact really exist. This notation, by reifying events and conditions, provides a way of specifying higher-order properties in first-order logic. This Davidsonian reification of eventualities (Davidson, 1967) is a common device in AI. Hobbs (1985) provides further explanation of the specific notation and ontological assumptions.

In this notation, the logical form of a sentence is a flat, scope-free, conjunction of positive literals, each of which is a predicate applied to the appropriate number of existentially quantified variables. For example, the logical form of the sentence

A man walked slowly.

is

$$(\exists\, x,e)\; man(x) \wedge past(e) \wedge walk'(e,x) \wedge slow(e)$$

2.2 Interpretation as Abduction

Abductive inference is inference to the best explanation. The process of interpreting sentences in discourse can be viewed as the process of providing the

---

[1] In the logical formulas in this paper, quantifiers are assumed to scope over logical operators.

best explanation of why the sentences would be true. This insight can be cashed out procedurally in terms of theorem-proving technology as follows:
To interpret a sentence:

> (1) Prove the logical form of the sentence,
>   including the constraints predicates impose on their arguments,
>   allowing for coercions,
>   Merging redundancies where possible,
>   Making assumptions where necessary.

In a discourse situation, the speaker and hearer both have their sets of private beliefs, and there is a large overlapping set of mutual beliefs. An utterance spans the boundary between mutual belief and the speaker's private beliefs. It is a bid to extend the area of mutual belief to include some private beliefs of the speaker's. It is anchored referentially in mutual belief, and where we succeed in proving the logical form and the constraints, we are recognizing this referential anchor. This is the given information, the definite, the presupposed. Where it is necessary to make assumptions, the information comes from the speaker's private beliefs, and hence is the new information, the indefinite, the asserted. Merging redundancies is a way of getting a minimal, and hence a best, interpretation.

Choosing the best or minimal interpretation relies on an algorithm for weighted abduction that levies variable costs for assumptions and for length of proof and reduces costs when redundancies are recognized. In IA the weighted abduction inference procedure, due to Stickel, is described in detail.
Consider a simple example.

The Boston office called.

This sentence poses at least three local pragmatics problems, the problems of resolving the definite reference of "the Boston office", expanding the metonymy to "*Some person at* the Boston office called", and determining the implicit relation between Boston and the office. Let us put these problems aside for the moment, however, and interpret the sentence according to characterization (1). We must prove abductively the logical form of the sentence together with the constraint "call" imposes on its agent, allowing for a coercion. That is, we must prove abductively the expression (ignoring tense and some other complexities)

$$(2)\ (\exists x,y,z,e)\ call'(e,x) \wedge person(x) \wedge rel(x,y) \wedge office(y) \wedge Boston(z) \wedge nn(z,y)$$

That is, there is a calling event e by x where x is a person. x may or may not be the same as the explicit subject y of the sentence, but x is at least related to y, or coercible from y, represented by $rel(x,y)$. y is an office and bears some unspecified relation nn to z, which is Boston. $person(x)$ is the requirement that $call'$ imposes on its agent x.

The sentence can be interpreted with respect to a knowledge base of mutual knowledge that contains the following facts:

$Boston(B_1)$

that is, B_1 is the city of Boston.

$\quad$ *office*(O_1) $\wedge$ *in*(O_1, B_1)

that is, O_1 is an office and is in Boston.

$\quad$ *person*(J_1)

that is, John J_1 is a person.

$\quad$ *work-for*(J_1, O_1)

that is, John J_1 works for the office O_1.

$\quad$ $(\forall y, z)$ *in*$(y, z) \supset$ *nn*(z, y)

that is, if y is in z, then z and y are in a possible compound nominal relation.

$\quad$ $(\forall x, y)$ *work–for*$(x, y) \supset$ *rel*(x, y)

that is, if x works for y, then y can be coerced into x.

The proof of all of (2) is straightforward except for the conjunct *call'*(x). Hence, we assume that; it is the new information conveyed by the sentence. This interpretation is illustrated in the proof graph of Figure 1, where a rectangle is drawn around the assumed literal *call'*(e, x).

Logical Form:

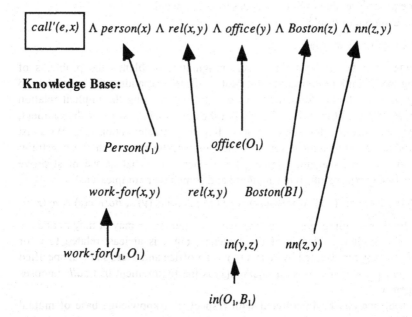

Figure 1. Interpretation of "The Boston office called."

Now notice that the three local pragmatics problems have been solved as a by-product. We have resolved "the Boston office" to O_1. We have determined the implicit relation in the compound nominal to be in. And we have expanded the metonymy to "John, who works for the Boston office, called."

In IA a number of other examples are presented showing how this approach yields solutions to problems of syntactic and lexical ambiguity, the resolution of pronouns and implicit arguments, the interpretation of compound nominals, the expansion of metonymies, and schema recognition. Hobbs (1992) uses the abductive approach to deal with metaphor. In IA it is shown how the interpretation as abduction approach can be combined with the parsing as deduction approach to yield a smooth integration of syntax, compositional semantics, and local pragmatics. It also sketches how this approach can be extended to the recognition of the structure of discourse.

The present paper is the beginning of an effort to extend the framework to global pragmatics, that is, to the recognition of the role of the discourse in the participants' ongoing plans.

2.3 The Form of Axioms

Because of the use of eventualities, often axioms that intuitively ought to be written as

$$(\forall x)\, p(x) \supset q(x)$$

will be written

$$(\forall e_1, x)\, p'(e_1, x) \supset (\exists e_2)\, q'(e_2, x)$$

That is, if e_1 is the eventuality of p being true of x, then there is an eventuality e_2 of q being true of x. It will sometimes be convenient to state this in a stronger form. It is not just that if e_1 exists, then e_2 happens to exist as well. The eventuality e_2 exists *by virtue of the fact that* e_1 exists. Let us express this tight connection by the predicate *gen*, for "generates". Then the above axiom can be strengthened to

$$(\forall e_1, x)\, p'(e_1, x) \supset (\exists e_2)\, q'(e_2, x)\, (gen(e_1, e_2))$$

Not only is there an e_2, but there an e_2 by virtue of the fact that there is an e_1. The relative existential and modal statuses of e_1 and e_2 can then be axiomatized in terms of the predicate *gen*.

In this paper, the predicates *cause*, *enable* and *imply* will sometimes play the role of *gen*.

It might seem in abduction that since we use only backchaining to find a proof and a set of assumptions, we cannot use superset information. However, the fact that we can make assumptions enables us to turn axioms around. In general, an axiom of the form

$$species \supset genus$$

can be converted into a biconditional axiom of the form

genus $\wedge$ differentiae $\supset$ species

Often we will not be able to prove the differentiae, and in many cases we cannot even spell them out. But in our abductive scheme, this does not matter; they can simply be assumed. In fact, we need not state them explicitly at all. We can simply introduce a predicate, a different one for each axiom, that stands for all the remaining properties. It will never be provable, but it will be assumable. Thus, in addition to having axioms like

$(\forall x)\ elephant(x) \supset mammal(x)$

we may have axioms like

$(\forall x)\ mammal(x) \wedge etc_1(x) \supset elephant(x)$

Then, even though we are strictly backchaining in search for an explanation, the fact that something is a mammal can still be used as (weak) evidence for its being an elephant, since we can assume the "et cetera" predication $etc_1(x)$ for a certain cost.

We can read this axiom as saying, "One way of being a mammal is being an elephant."

This device may seem ad hoc at first blush. But I view the device as implementing a fairly general solution to the problems of nonmonotonicity in commonsense reasoning and vagueness of meaning in natural language, very similar to the use of abnormality predicates in circumscriptive logic (McCarthy, 1987). Whereas, in circumscriptive logic, one typically specificies a partial ordering of abnormality predicates in accordance with which they are minimized, in the weighted abduction framework, one uses a somewhat more flexible system of costs.

There is no particular difficulty in specifying a semantics for the "et cetera" predicates. Formally, etc1 in the axiom above can be taken to denote the set of all things that are either not mammals or are elephants. Intuitively, etc1 conveys all the information one would need to know beyond mammalhood to conclude something is an elephant. As with nearly every predicate in an axiomatization of commonsense knowledge, it is hopeless to spell out necessary and sufficient conditions for an "et cetera" predicate. In fact, the use of such predicates in general is due largely to a recognition of this fact about commonsense knowledge.

The "et cetera" predicates constitute one of the principal devices for giving our logic "soft corners". We would expect them to pervade the knowledge base. Virtually any time there is an axiom relating a species to a genus, there should be a corresponding axiom, incorporating an "et cetera" predication, expressing the inverse relation.

Let us summarize at this point the most elaborate form axioms in the knowledge base will have. If we wish to express an implicative relation between concepts p and q, the most natural way to do so is as the axiom

$$(\forall x,z)\, p(x,z) \supset (\exists y)\, q(x,y)$$

where z and y stand for arguments that occur in one predication but not in the other. When we introduce eventualities, this axiom becomes

$$(\forall e_1,x,z)\, p'(e_1,x,z) \supset (\exists e_2,y)\, q'(e_2,x,y)$$

Using the *gen* relation to express the tight connection between the two eventualities, the axiom becomes

$$(\forall e_1,x,z)\, p'(e_1,x,z) \supset (\exists e_2,y)\, q'(e_2,x,y) \wedge gen(e_1,e_2)$$

Next we introduce an "et cetera" proposition into the antecedent to take care of the imprecision of our knowledge of the implicative relation.

$$(\forall e_1,x,z)p'(e_1,x,z) \wedge etc_1(x,z) \supset (\exists e_2,y)q'(e_2,x,y) \wedge gen(e_1,e_2)$$

Finally we biconditionalize the relation between p and q by writing the converse axiom as well:

$$(\forall e_1,x,z)p'(e_1,x,z) \wedge etc_1(x,z) \supset (\exists e_2,y)q'(e_2,x,y) \wedge gen(e_1,e_2)$$

$$(\forall e_1,x,y)q'(e_2,x,y) \wedge etc_2(x,y) \supset (\exists e_1,z)p'(e_1,x,z) \wedge gen(e_2,e_1)$$

This then is the most general formal expression in our abductive logic of what is intuitively felt to be an *association* between the concepts p and q.

In this paper, for notational convenience, I will use the simplest form of axiom I can get away with for the example. The reader should keep in mind however that these are only abbreviations for the full, biconditionalized form of the axiom.

3. An Example of Plan Recognition

3.1 The Example

Let us analyze an example from a set of dialogues collected by Barbara Grosz (1977) between an expert and an apprentice engaged in fixing an air compressor. They are in different rooms, communicating by terminals. The apprentice A is doing the actual repairs, after receiving instructions from the expert B. At one point, the following exchange takes place:

B: Tighten the bolt with a ratchet wrench.
A: What's a ratchet wrench?
B: It's between the wheel puller and the box wrenches.

A seems to be asking for a definition of a ratchet wrench. But that is not what B gives her. He does not say

A ratchet wrench is a wrench with a pawl, or hinged catch, that engages the sloping teeth of a gear, permitting motion in one direction only.

Instead he tells her where it is.

According to a plausible analysis, B has interpreted A's utterance by relating it to A's overall plan. B knows that A wants to use the ratchet wrench. To use a ratchet wrench, you have to know where it is.

To know where it is, you have to know what it is. B responds to A's question, not by answering it directly, but by answering to a higher goal in A's presumed overall plan, by telling A where it is.

B has therefore recognized the relationship between A's utterance and her overall plan. I will give two accounts of how this recognition could have taken place. The first account is informational. It is derived in the process of proving the logical form. The second account is intentional and subsumes the first. It is derived in the process of explaining, or proving abductively, the fact that A's utterance occurred.

3.2 The Informational Solution

For this solution we will need two axioms encoding the planning process:

(3) $(\forall a, e_0, e_1)\, goal(a, e_1) \wedge enable(e_0, e_1) \supset goal(a, e_0)$

or if an agent a has e_1 as a goal and e_0 enables, or is a prerequisite for, e_1, then $a\$$ has e_0 as a goal as well.

(4) $(\forall a, e_0, e_1)\, goal(a, e_1) \wedge cause(e_0, e_1) \wedge etc_1(a, e_0, e_1) \supset goal(a, e_0)$

or if an agent a has e_1 as a goal and e_0 causes, or is one way to accomplish, e_1, then a may have e_0 as a goal as well. The etc_1 literal encodes the uncertainty as to whether e_0 will be chosen as the way to bring about e_1 rather than some other action that causes e_1.

In terms of STRIPS operators (Fikes and Nilsson, 1971), the first axiom says that prerequisites for an action must be satisfied, while the second axiom says essentially that to achieve a goal, an operator needs to be chosen and its body (e_0) needs to be executed.

Next we need two domain axioms of a rather general character.

(5) $(\forall e_2, a, x)\, use'(e_2, a, x) \supset (\exists e_3, e_4, y) enable(e_3, e_2) \wedge know'(e_3, a, e_4) \wedge at'(e_4, x, y)$

or an agent a's use e_2 of a thing x has as a prerequisite a's knowing e_3 the fact e_4 that x is at someplace y. To use something, you have to know where it is.

(6) $(\forall e_3, a, e_4, x, y)\, know'(e_3, a, e_4) \wedge at'(e_4, x, y)$
$\quad \supset (\exists e_5, e_6) enable(e_5, e_3) \wedge know'(e_5, a, e_6) \wedge wh'(e_6, x)$

or an agent a's knowing e_3 the fact e_4 that a thing x is at someplace y has as a prerequisite a's knowing e_5 what x is (e_6). To know where something is, you have to know what it is. We dodge the complex problem of specifying what constitutes knowing what something is by encoding it in the predicate wh, which represents the relevant context-dependent essential property.

Let us suppose that the logical form of

What's a ratchet wrench?

is

(7) $(\exists a,e_5,e_6)\ goal(a,e_5)\ \land\ know'(e_5,a,e_6)\ \land\ wh'(e_6,RW)$

That is, the speaker a has the goal e_5 of knowing the essential property e_6 of the ratchet wrench RW. Most of this logical form comes, of course, from our recognition that the utterance is a question.

Suppose also that in B's knowledge of the context is the following fact:

(8) $goal(A,E_2)\ \land\ use'(E_2,A,RW)$

That is, the apprentice A has the goal E_2 of using the ratchet wrench RW.

The proof of the logical form (7) follows from axioms (3) through (6) together with fact (8), as indicated in Figure 2. Axiom (3) is used twice, first in conjunction with axiom (6) and then with axiom (5), to move up the planning tree. The apprentice wants to know what a ratchet wrench is because she wants to know where it is, and she wants to know where it is because she wants to use it. The proof then bottoms out in fact (8).

Logical Form:

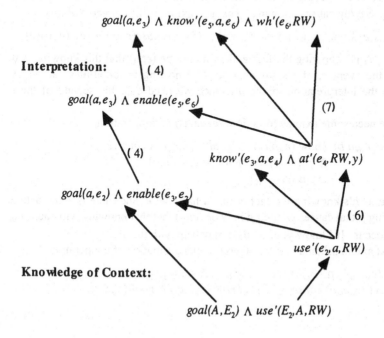

Figure 2. Informational interpretation of "What's a ratchet wrench?"

To summarize, if we take the logical form of a question to be the expression of a desire to know something, then the proof of that logical form often involves the recognition of the ultimate aims of the speaker in asking it.

3.3 The Intentional Solution

According to the Informational Perspective, it is the logical form of the utterance that needs to be explained, or proven abductively. We will now take a broader view in which it is the occurrence of an event in the world that has to be explained. It is not the content of the utterance that we have to explain, but rather the very fact that the utterance occurred. Frequently, the best explanation of an event is that it is an intentional action on the part of some agent, that is, it is an action in the service of some goal. This is especially true of utterances—they are generally intentional acts. Thus, we will be interpreting the utterance from an Intentional Perspective. We will ask why the speaker said what she did. We will see how this in turn encompasses the Informational Perspective.

We need several more axioms. First we need some axioms about speaking.

$$(\forall e_7, a, b, e_8)\ say'(e_7, a, b, e_8) \supset (\exists e_9)\ cause(e_7, e_9) \wedge know'(e_9, b, e_8)$$

That is, if e_7 is a's saying e_8 to b, then that will cause the condition e_9 of b's knowing e_8. Saying causes knowing. The next axiom is the converse of this.

$$(9)\ (\forall e_k, y, e)\ know'(e_k, y, e) \wedge etc_2(e_k, y, e) \supset (\exists e_s, x)\ cause(e_s, e_k) \wedge say'(e_s, x, y, e)$$

That is, if e_k is y's knowing the fact e, then it may be (etc_2) that this knowing was caused by the event e_s of x's saying e to y. Knowing is sometimes caused by saying. In the interpretation of the utterance we need only the second of these axioms.

Next we need some axioms (or axiom schemas) of cooperation.

$$(\forall e_5, e_8, e_9, e_{10}, a, b)\ know'(e_9, b, e_8) \wedge goal'(e_8, a, e_5) \wedge cause(e_{10}, e_5)$$
$$\wedge\ p'(e_{10}, b) \wedge etc_3(e_5, e_8, e_9, e_{10}, a, b)$$
$$\supset cause(e_9, e_{10})$$

That is, if e_9 is b's knowing the fact e_8 that a has goal e_5 and there is some action e_{10} by b doing p that causes e_5, then it may be (etc_3) that that knowing will cause e_{10} to actually occur. If I know your goals, I may help with them[2].

The next axiom schema is the converse, being attribution of cooperation.

$$(10)\ (\forall e_5, e_{10}, b)\ p'(e_{10}, b) \wedge cause(e_{10}, e_5) \wedge etc_4(e_5, e_{10}, b)$$
$$\supset (\exists e_8, e_9, a)\ cause(e_9, e_{10}) \wedge know'(e_9, b, e_8) \wedge goal'(e_8, a, e_5)$$

---

[2] More properly, where I have $p'(e_{10}, b)$ I should have $agent(b, e_{10})$, together with a set of axioms of the form $(\forall e, x)\ p'(e, x, ...) \supset agent(x, e)$.

That is, if an action e_{10} by b occurs, where e_{10} can cause e_5, then it may be (*etc*$_4$) that it was caused by the condition e_9 of b's knowing the fact e_8 that a has the goal e_5. Sometimes I do things because I know it will help you. In the example we will only need the axiom in this direction.

Finally, an axiom schema that says that people do what they want to do.

(11) $(\forall a,e_7)\ goal(a,e_7) \wedge p'(e_7,a) \wedge etc_5(a,e_7) \supset Rexists(e_7)$

That is, if a has as a goal some action e_7 that a can perform, then it could be (*etc*$_5$) that e_7 will actually occur. This axiom, used in backward chaining, allows us to attribute intention to events.

Now the problem we set for ourselves is not to prove the logical form of the utterance, but rather to explain, or prove abductively, the occurrence of an utterance with that particular content. We need to prove

$$(\exists e_7,a,b,e_8,e_5,e_6)\ Rexists(e_7) \wedge say'(e_7,a,b,e_8)$$
$$\wedge\ goal'(e_8,a,e_5) \wedge know'(e_5,a,e_6) \wedge wh'(e_6,RW)$$

That is, we need to explain the existence in the real world of the event e_7 of someone a saying to someone b the proposition e_8 that a has the goal e_5 of knowing the essential property e_6 of a ratchet wrench.

The proof of this is illustrated in Figure 3. The boxes around the "et cetera" literals indicate that they have to be assumed. By axiom (11) we attribute intention to explain the occurrence of the utterance act e_7; it's not like a sneeze. Using axiom (4), we hypothesize that this intention or goal is a subgoal of some other goal e_9. Using axiom (9), we hypothesize that this other goal is b's knowing the content e_8 of the utterance. A uttered the sentence so that B would know its content. Using axiom (4) again, we hypothesize that e_9 is a subgoal of some other goal e_{10}, and using axiom (10) we hypothesize that e_{10} is b's saying e_6 to a. A told B A's goal so that B would satisfy it. Using axiom (4) and (9) again, we hypothesize that e_{10} is a subgoal of e_5, which is a's knowing e_6, the essential property of a ratchet wrench. A wants B to tell her what a ratchet wrench is so she will know it.

The desired causal chain is this: A tells B she wants to know what a ratchet wrench is, so B will know that she wants to know what a ratchet wrench is, so B will tell her what a ratchet wrench is, so she will know what a ratchet wrench is. Causal chains are reversed in planning; if X causes Y, then our wanting Y causes us to want X. Hence, the causal chain is found by following the arrows in the diagram in the reverse direction.

At this point all that remains to prove is

$$(\exists\ a,e_5,e_6)\ goal(a,e_5) \wedge know'(e_5,a,e_6) \wedge wh'(e_6,RW)$$

But this is exactly the logical form whose proof is illustrated in Figure 2. We have reduced the problem of explaining the occurrence of an utterance to the problem of discovering its intention, and then reduced that to the problem of explaining the content of the utterance. Interpetation from the Intentional Perspective includes as a subpart the interpretation of the utterance from the Informational Perspective.

Observable to be Explained:

$Rexists(e_7) \wedge say'(e_7, a, b, e_8) \wedge goal'(e_8, a, e_5)$

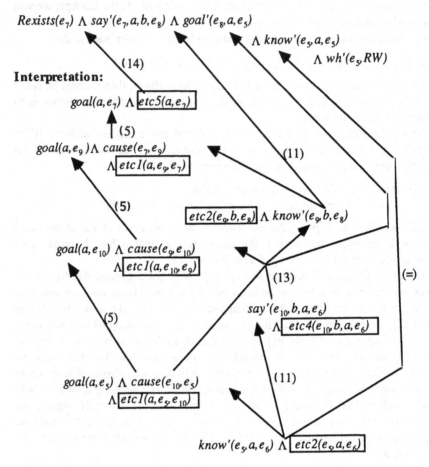

Figure 3. Intentional Interpretation of "What's a ratchet wrench?"

4. Tautology

The framework that has been presented here gives us a handle on some of the more complex things speakers do with their utterances. Let us see how we could deal with one example—tautology, such as "boys will be boys," "fair is fair," and "a job is a job." Norvig and Wilensky (1990) cite this figure of speech as something that should cause trouble for an abduction approach that seeks minimal explanations, since the minimal explanation is that they just express a known truth. Such an explanation requires no assumptions at all.

In fact, the phenomenon is a good example of why an informational account of discourse interpretation has to be embedded in an intentional account. Let us imagine two mothers, A and B, sitting in the playground and talking.

A: Your Johnny is certainly acting up today, isn't he?

B: Boys will be boys.

In order to avoid dealing with the complications of plurals and tense in this example, let us simplify B's utterance to

B: A boy is a boy.

From the Informational Perspective, several interpretations of this utterance are possible. The first is the Literal Extensional Interpretation. The first "a boy" introduces a specific, previously unidentified boy and the second says about him that he is a boy. The second is the Literal Intensional Interpretation. The sentence expresses a trivial implicative relation between two general propositions—$boy(x)$ and $boy(x)$. The third is the Desired Interpretation. The first "a boy" identifies the typical member of a class which Johnny is a member of and the second conveys a general property, "being a boy", as a way of conveying a specific property, "misbehaving", which is true of members of that class.

More precisely, the logical form of the sentence can be written as follows:

$$(\exists e_1, e_2, x, y, z, w)\ boy'(e_1, x) \wedge rel(z, x) \wedge be(z, w) \wedge rel(w, y) \wedge boy'(e_2, y)$$

This sentence expresses a be relation between two entities, but either or both of its arguments may be subject to coercion. Thus, we have introduced the two rel relations. The logical form can be given the tortured paraphrase, "z is w, where z is related to x whose boy-ness is e_1 and w to y whose boy-ness is e_2."

The required axioms are as follows:

Everything is itself:

$$(\forall x)\ be(x, x)$$

Implication can be expressed by "to be":

$$(\forall e_1, e_2)\ imply(e_1, e_2) \supset be(e_1, e_2)$$

Implication is reflexive:

$$(\forall e)\ imply(e, e)$$

Boys misbehave:

$$(\forall e_1, x)\ boy'(e_1, x) \supset (\exists e_3)\ misbehave'(e_3, x)$$

Misbehavers are often boys:

$$(\forall e_3, x)\ misbehave'(e_3, x) \wedge etc_1(x) \supset (\exists e_2)\ boy'(e_2, x)$$

Identity is a possible coercion relation:

$$(\forall x)\ rel(x, x)$$

An entity can be coerced into a property of the entity:

$$(\forall e,x)\ boy'(e,x) \supset rel(e,x)\ (\forall e,x)\ misbehave'(e,x) \supset rel(e,x)$$

Now the Literal Extensional Interpretation is established by taking the two coercion relations to be identity, taking *be* to be expressing identity, and assuming $boy(e_1,x)$ (or equivalently, $boy(e_2,y)$).

In the Literal Intensional Interpretation, z is identified with e_1, w is identified with e_2, and $boy'(e_1,x)$ and $boy'(e_2,y)$ are taken to be the two coercion relations. Then e_2 is identified with e_1 and $be(e_1,e_1)$ is interpreted as a consequence of $imply(e_1,e_1)$. Again, $boy(e_1,x)$ is assumed.

In the Desired Interpretation, the first coercion relation is taken to be $boy'(e_1,x)$, identifying z as e_1. The second coercion relation is taken to be $misbehave'(e_3,y)$, identifying w as e_3. If $etc_1(y)$ is assumed, then $misbehave'(e_3,y)$ explains $boy(e_2,y)$. If $boy(e_1,x)$ is assumed, it can explain $misbehave'(e_3,y)$, identifying x and y, and also $imply(e_1,e_3)$. The latter explains $be(e_1,e_3)$.

From the Informational Perspective alone, the Literal Extensional Interpretation is minimal and hence would be favored. The Desired Interpretation is the worst of the three.

But the Literal Extensional and Intensional Interpretations leave the *fact* of the utterance unaccounted for. From the Intentional Perspective, this is what we need to explain. The explanation would run something like this:

B wants A to believe that B is not responsible for Johnny's misbehaving.

Thus, B wants A to believe that Johnny misbehaves necessarily.

Thus, given that Johnny is necessarily a boy, B wants A to believe that Johnny's being a boy implies that he misbehaves.

Thus, B wants to convey to A that being a boy implies misbehaving.

Thus, given that boy-ness implies misbehaving is a possible interpretation of a boy being a boy, B wants to say to A that a boy is a boy.

The content of the utterance under the Literal Extensional and Intensional Interpretations do not lend themselves to explanations for that fact of the utterance in the way that the Desired Interpretation does. The requirement for the *globally* minimal explanation in the Intentional Perspective, that is, the requirement that both the content and the fact of the utterance must be explained, forces us into an interpretation of the content that would not be favored from the Informational Perspective alone.

The two literal interpretations give us good explanations for the content of the sentence but do not give us good explanations for the saying of a sentence with that content. The Desired Interpretation, however, does fit into an explanation of why the utterance was uttered. It can be paraphrased as "Members of the class that Johnny belongs to always behave in this fashion," and it thus defends B against the implied accusation that she is not a good mother.

We are forced into an interpretation of the content that, while not optimal locally, contributes to a global interpretation that *is* optimal.

5. Summary

For a discourse consisting of assertions from a speaker A to a hearer B,[3] the relation between the intentional account of interpretation and the informational account can be summarized succinctly by the following formula:

(12) **intentional-account** = *goal*(A, *believe*(B, **informational-account**))

The speaker ostensibly has the goal of changing the beliefs of the hearer to include the content characterized by the informational account. When we reason about the speaker's intention, we are reasoning about how this goal fits into the larger picture of her ongoing plan. We are asking why she seems to be trying to get the hearer to believe this content.

In some cases there is a strong correspondence between the two accounts. The content is something that is reasonable to believe and it is easy to see why the speaker wants the hearer to believe it in the given situation. In other cases, there is very little information about one account or the other. In pragmatically elliptical utterances, the informational account is highly underdetermined and the global interpretation is thus primarily shaped by the intentional account. In beginnings, chance encounters with strangers, and random remarks, the hearer can guess very little about the intention of the speaker except via the informational account, so that is primarily what shapes the global interpretation. Finally, there are cases of genuine conflict between the two accounts. The informational account does not seem to be true, or it seems to run counter to the speaker's goals for the hearer to come to believe it, or it ought to be obvious that the hearer already does believe it. Tautologies are an example of the last of these cases. In these cases, the intentional account may force an alternate, ordinarily nonoptimal informational reading, or the hearer may be forced to reassess the speaker's goals.

The framework presented in IA gives a uniform account of how one determines the informational interpretation of a sentence or larger stretch of text, in terms of abductive inference. The present paper begins to extend that framework to encompass intentional interpretation as well.

Schema (12) points to a third major area of research, beyond the determination of intention and information—the problem of belief revision. Given that the speaker's goal is to get the hearer to believe the informational reading of the utterance, under what circumstances should the hearer actually come to believe it. This, however, is an issue for another paper.

---

[3] Similar schemas could be given for questions and orders.

Acknowledgments

The research described here was supported by the National Science Foundation and the Advanced Research Projects Agency under Grant IRI-9314961 (Integrated Techniques for Generation and Interpretation).

References

Allen, J.F. and C.R. Perrault. 1980. Analyzing Intention in Utterances. *Artificial Intelligence* 15, 143–178.

Bruce, B.C. 1975. Belief Systems and Language Understanding. Technical Report 2973, Bolt, Beranek, and Newman, Cambridge, Massachusetts.

Cohen, P. and C.R. Perrault. 1979. Elements of a Plan-based Theory of Speech Acts. *Cognitive Science* 3(3), 177–212.

Davidson, D. 1967. The Logical Form of Action Sentences. In N. Rescher (ed) *The Logic of Decision and Action*. University of Pittsburgh Press, Pittsburgh, Pennsylvania. 81–95.

Fikes, R. and N.J. Nilsson. 1971. STRIPS: A New Approach to the Application of Theorem Proving to Problem Solving. *Artificial Intelligence* 2, 189–208.

Grosz, B.J. 1977. The Representation and Use of Focus in Dialogue Understanding. Stanford Research Institute Technical Note 151, Stanford Research Institute, Menlo Park, California, July 1977.

Grosz, B.J. 1979. Utterance and Objective: Issues in Natural Language Communication. *Proceedings of the 6th International Joint Conference on Artificial Intelligence* IJCAI, Tokyo, Japan. 1067–1076.

Grosz, B.J. and C. Sidner. 1986. Attention, Intentions, and the Structure of Discourse. *Computational Linguistics* 12(3), 175–204.

Hobbs, J.R. 1983. An Improper Treatment of Quantification in Ordinary English. *Proceedings of the 21st Annual Meeting, Association for Computational Linguistics*. Cambridge, Massachusetts. 57–63.

Hobbs, J.R. 1985. Ontological Promiscuity. *Proceedings of the 23rd Annual Meeting of the Association for Computational Linguistics*. Chicago, Illinois. 61–69.

Hobbs, J.R. 1992. Metaphor and Abduction. In A. Ortony, J. Slack, and O. Stock (eds) *Communication from an Artificial Intelligence Perspective: Theoretical and Applied Issues*. Berlin: Springer-Verlag. 35–58. Also published as SRI Technical Note 508, SRI International, Menlo Park, California. August 1991.

Hobbs, J.R. 1995. Monotone Decreasing Quantifiers in a Scope-Free Logical Form. In K. van Deemter and S. Peters (eds), *Semantic Ambiguity and Underspe-cification*. CSLI Lecture Notes, Stanford, California.

Hobbs, J.R. and D.A. Evans. 1980. Conversation as Planned Behavior. *Cognitive Science* 4(4), 349–377.

Hobbs, J.R., M. Stickel, D. Appelt, and P. Martin. 1993. Interpretation as Abduction. *Artificial Intelligence* (to appear). Also published as SRI Technical Note 499, SRI International, Menlo Park, California. December 1990.

McCarthy, J. 1987. Circumscription: A Form of Nonmonotonic Reasoning. In M. Ginsberg (ed) *Readings in Nonmonotonic Reasoning*. Los Altos CA: Morgan Kaufmann, 145–152.

Norvig, P. and R. Wilensky. 1990.   A Critical Evaluation of Commensurable Abduction Models for Semantic Interpretation.   In H. Karlgren (ed) *Proceedings of the 13th International Conference on Computational Linguistics COLING* Vol 3, Helsinki, Finland. 225–230.

Perrault, C.R. and J.F. Allen. 1980.   A Plan-Based Analysis of Indirect Speech Acts. *American Journal of Computational Linguistics* 6(3–4), 167–182.

Power, R. 1974.   A Computer Model of Conversation.   Ph.D. dissertation, University of Edinburgh, Scotland.

Schmidt, C.F., N.S. Sridharan, and J.L. Goodson. 1978.   The Plan Recognition Problem: An Intersection of Psychology and Artificial Intelligence. *Artificial Intelligence* 11, 45–83.

An Empirical Perspective

Chapter 7
Empirical Analysis of Three Dimensions of Spoken Discourse: Segmentation, Coherence, and Linguistic Devices

Rebecca J. Passonneau[1] and Diane J. Litman[2]

[1] Columbia University, New York

[2] AT&T Bell Laboratories, Murray Hill

1. Discourse Segmentation

A discourse consists not simply of a linear sequence of utterances,[1] but of meaningful semantic or pragmatic relations among utterances. Each utterance of a discourse either bears a relation to a preceding utterance or constitutes the onset of a new unit of meaning or action that subsequent utterances may add to. The need to model the relation between such units and linguistic features of utterances is almost universally acknowledged in the literature on discourse. For example, previous work argues for an interdependence between particular cue words and phrases such as anyway, and their location relative to an utterance or text, (e.g., Hirschberg and Litman 1993, Grosz and Sidner 1986, Reichman 1985, Cohen 1984); the distribution and duration of pauses relative to multi-utterance units (e.g., Grosz and Hirschberg 1992, Hirschberg and Grosz 1992, Chafe 1980, Butterworth 1980); and the interdependence between the form of discourse anaphoric noun phrases and the relation of the current utterance to a hierarchical model of utterance actions, or to a model of focus of attention (e.g., Grosz 1977, Grosz and Sidner 1986, Reichman 1985, Sidner 1979, Passonneau 1985). However, there are a variety of distinct proposals regarding how to model the interdependence among the three dimensions of: 1. sequences of semantically and pragmatically related utterances, 2. the units or relations they reflect, and 3. lexico-grammatical or prosodic features. We refer to these dimensions respectively as segmentation, coherence, and linguistic devices. For purposes of this paper we have deliberately avoided adopting a particular theoretical framework in order to pose open-ended questions that we address through an empirical investigation.

Figure 1 presents an excerpt from a transcription of a spoken narrative taken from the Pear stories corpus (Chafe 1980) that we use in our study. We use the excerpt to illustrate the interdependence of segmentation, linguistic devices, and

---

[1] We use the term *utterance* to mean a use of a sentence or other linguistic unit, whether in text or in spoken language.

coherence. Chafe's transcriptions (partly modified for Figure 1) represent all lexical articulations including repeated and incomplete words and phrases; non-lexical articulations such as *uh, um, tsk*; vowel lengthening as indicated by '-'; location and duration of pauses in seconds (cf. numbers in brackets)[2]; and prosodic phrase units. A period or question mark at the end of a line indicates the end of a prosodic unit with utterance final intonation (i.e., assertion or question) and a comma at the end of a line indicates the end of a more intermediate prosodic unit (roughly corresponding to the distinction between intermediate and complete prosodic phrases in (Beckman 1991) and (Pierrehumbert 1980)). We have numbered the prosodic phrases sequentially, with the second field corresponding to the intermediate units within complete prosodic phrases.

We have grouped together certain sequences of utterances and given them segment labels, showing only the first two utterances of segment 3. Three of the many linguistic devices which correlate with this segmental structure are pauses, cue words, and referential noun phrases. First, long pauses consisting of silence or of non-lexical articulations (e.g., throat clearing) occur before the first proposition of all 3 segments. Second, assuming that *a-nd* in 9.1 is a cue word signalling a relation of 9.1 to the discourse, (rather than part of the proposition conveyed by *he goes up the ladder*), then 2 segments (1 and 3) begin with the cue words *okay* and *a-nd*. Finally, the discourse anaphoric pronoun *he* in 9.1 is used to refer to an entity first introduced in segment 1 as *a farmer* and later referred to as *he* in every remaining utterance of the segment. Despite the subsequent introduction of another male entity in segment 2, *a man with a goat*, the reference of the underlined pronoun in segment 3 is not ambiguous between the two male entities. This in part reflects differences in the semantic content of segments 1 and 2, and the close semantic relation of 9.1 to segment 1. It illustrates a phenomenon referred to as a discourse pop (Grosz 1977, Reichman 1985, Grosz and Sidner 1986, Fox 1987) in which a suspended discourse goal, e.g., describing the farmer, is later resumed. The indentation in Figure 1 thus reflects that segment 3 resumes the communicative goal suspended after segment 1, whereas segment 2 addresses a distinct communicative goal. Our investigations are directed at determining what regularities there are in the interaction of segmentation, coherence and the three types of linguistic devices mentioned here: pauses, cue words and referential noun phrases.

While there is strong consensus on the need to model one or more aspects of the interaction among segmentation, coherence and linguistic devices, there is either weak consensus or only preliminary research on many fundamental issues.

---

[2] For pairs of brackets containing pauses interspersed with other material, the outermost duration indicates the duration of the whole sequence of pauses; e.g., '[4.4 [.8 sniff [2.5] *throat clearing*]' indicates a 4.4 second duration beginning with a .8 second pause, followed by a sniff, and so on.

SEGMENT 1

1.1   Okay.

2.1   [4.4 [.8 sniff [2.5] throat clearing] [1.55] tsk [.35]
       There's [.65] a farmer,
2.2   [.15] he looks like ay uh Chicano American,
2.3   [.5] he is picking pears.

3.1   [4.85 [2.5] A-nd u-m [3.35]] he's just picking them,
3.2   he comes off of the ladder,
3.3   [3.5+ [.35] a-nd] he- u-h [.3] puts his pears into the basket.

SEGMENT 2

4.1   [1.0 [.5] U-h] a number of people are going by,
4.2   [.35+ and [.35]] one is [1.15 um] /you know/ I don't know,
4.3   I can't remember the first...the first person that goes by.

5.1   [.3] Oh.

6.1   A u-m a man with a goat [.2] comes by.

7.1   [.25] It see it seems to be a busy place.

8.1   [.1] You know,
8.2   fairly busy,
8.3   it's out in the country,
8.4   [.4] maybe in u-m [.8] u-h the valley or something.

SEGMENT 3

9.1   [2.95 [.9] A-nd um [.25] throat clearing [.35]] he goes up the ladder,
9.2   and picks some more pears.

       ...

Figure 1. Discourse Segment Structure: A Typical 'Pop'

In Section 2, we present an overview of various models of the three dimensions of discourse we address, with a particular emphasis on attempts at empirical verification. Then we present interim results of our ongoing empirical study of segmentation of narrative discourses from Chafe's (1980) pear story corpus. We address the following questions:

- are discourse segments objective units that correspond directly to more abstract semantic or pragmatic units?
- how can theories of discourse segmentation be empirically verified?
- what is the nature of the boundaries between segments; e.g., do they have precise locations?
- what linguistic devices correlate with segment boundaries, and to what degree?

This chapter presents our take on the questions itemized above. Because the questions are interrelated, each section of the paper contributes gradually to our evolving view. In Section 3, we present our empirical methodology and the results of an analysis of segmentation data we collected from a population of naive subjects. The segmentation criterion was based on a view of communication as a type of action (cf. Austin 1962), hence related to intentionality in the sense of Grosz and Sidner (1986). Our results, consisting of a quantitative evaluation of the degree of reliability among subjects, demonstrate an extremely significant pattern of agreement on segment boundaries. However, we also point to a large amount of variability in the data, including imprecision in the location of certain boundaries and differences in what we refer to as boundary strength. In Section 4, we quantitatively evaluate the performance of three algorithms based on referring noun phrases, cue phrases, and pauses. Our ultimate goal is to determine whether the correlations are strong enough to be useful in NLP systems, a question we return to in the concluding section. Our results suggest that it would be possible to algorithmically achieve human levels of performance given multiple knowledge sources. However, due to variability within and across speakers, we believe the most effective algorithm would need to dynamically adjust to different cues, possibly dependent on user modelling. In Section 5 we relate our empirical results to coherence. First, we present a qualitative assessment of how well subjects agree on their abstractions of what segments are about. Then, we look in detail at an episode that occurs in all 10 narratives in our corpus, but which varies in size from 1 to 12 segments. This gives us one context in which to examine the relationship among segments, segment boundaries, and supra-segmental coherence. One of the algorithms is re-evaluated on this episode. Finally, we examine discourse pops, an example of another type of supra-segmental coherence. In the conclusion we discuss implications for natural language understanding and generation systems.

2. Related Work

We do not have space here to provide a complete review of existing claims regarding the interdependence of segmentation, coherence, and linguistic devices. Instead, we mention certain works in order to illustrate these points. First, many have argued that understanding or generating coherent discourse depends on an appropriate model of the relation among segmentation, coherence and linguistic devices. Second, few explicit claims are made regarding segmentation, but it implicitly plays a role in much work on discourse comprehension and generation. Third, semantic and pragmatic coherence among utterances subsumes both rhetorical and intentional relations.

Claims regarding segmental structure are most explicit in the work of Grosz and Sidner (1986), who propose a tri-partite discourse structure of three homomorphic levels: linguistic structure, which is roughly equivalent to what we refer to as segmental structure (but including linguistic devices), intentional structure, reflecting the goals of the discourse participants, and attentional state. They make the strong claim that these three levels are isomorphic. Relations among segments, and among focus spaces, are analogous to hierarchical relations among discourse segment purposes (DSP), or the intentions underlying individual segments. The significant relations among segments/DSPs/focus spaces pertain to whether whether one DSP *dominates* or *precedes* another, not to the specific types of relations among DSPs.

In much other work on discourse, segmental structure is an artifact of coherence relations among utterances, and few if any specific claims are made regarding segmental structure per se. We use coherence here in a general way to refer to semantic, lexical, referential, intentional, rhetorical, or other relations among utterances. For Hobbs (1979), understanding a discourse depends on understanding a variety of semantic and intentional coherence relations among utterances. Over time, he has made use of about half a dozen such relations, such as ELABORATION, EXPLANATION, OCCASION, and so on. He accounts for many other aspects of discourse coherence, such as assigning referents to noun phrases, as a by-product of understanding these inter-utterance relations. Polyani (1988) similarly lays out a specific set of relations among utterances, but comes closer to Grosz and Sidner in proposing a mapping from semantic and pragmatic relations among utterances to a tree structure of discourse constituent units (DCUs). As do Reichman (1985) and Fox (1987), Polanyi draws on work in interactional analysis (Schegloff and Sacks 1973) to distinguish between *open* or *closed* DCUs, and to predict the status of DCUs relative to the discourse parse tree. Open DCUs are located only on the right frontier of the tree, corresponding to the current focus stack of (Grosz and Sidner 1986) (cf. discussion in (Webber 1991)). Also, as in Grosz and Sidner, accessibility of discourse entities is predicted directly by the structure of the discourse tree.

Another tradition of defining relations among utterances arises out of Rhetorical Structure Theory (Mann and Thompson 1988), an approach which informs much work in generation (cf. discussion in Moore and Paris 1993). Moore and Paris argue that a generation system must have access to an explicit representation of intentions as well as rhetorical relations in order to handle phenomena such as clarifications of previous explanations. In related work, Moore and Pollack (1992) note problems that arise out of an assumption that a single RST relation must represent the relation among consecutive discourse elements. They attempt to disentangle the intentional from the informational relations in RST. While others (cf. Grosz and Sidner 1986, Grosz and Hirschberg 1992, Mann and Thompson 1988) have pointed out that the relation of an utterance to the global discourse model may be ambiguous, Moore and Pollack argue that the relation between intentional and informational structure is in some cases not isomorphic. They present an example of an utterance that is the nucleus of a set of utterances at the intentional level (e.g., a dominating DSP in terminology of Grosz and Sidner), but is not the nucleus when the same set of utterances is viewed on the informational level.

Although Moore and Pollack (1992) do not directly address the question of segmentation, their claim that an utterance can simultaneously be a nucleus (dominating) and a satellite (subordinate) raises a number of questions about the adequacy of trees or stacks as the appropriate means of representating global structure in discourse, and of the ability to predict constraints between this structure and linguistic devices. Although all of the studies cited above have carried out detailed analyses of individual discourses or representative corpora to arrive at more or less substantive hypotheses, we believe that there is a need for many more rigorous empirical analyses of discourse corpora of various modalities (e.g., spoken, written), genres (e.g., task oriented, narrative), registers (e.g., conversational, formal), and so on. In the remainder of this section, we discuss recent empirical work on the relation among segmentation, coherence and linguistic devices.

Recently, more rigorous empirical work has been directed at establishing objectively verifiable segment boundaries. Despite the observation that segmentation is rather subjective (Mann et al. 1992, Johnson 1985), several researchers have begun to investigate the ability of humans to agree with one another on segmentation, and to propose methodologies for quantifying their findings. Grosz and Hirschberg (Grosz and Hirschberg 1992, Hirschberg and Grosz 1992) asked subjects to structure three Associated Press news stories (averaging 450 words in length) according to the model of Grosz and Sidner (1986). Subjects identified hierarchical structures of discourse segments, as well as local structural features, using text alone as well as text and professionally recorded speech. Agreement ranged from 74%–95%, depending upon discourse feature. Hearst (1993a) asked subjects to place boundaries between paragraphs of three expository texts (length 77 to 160 sentences), to indicate topic changes. She found agreement greater than 80%. In (Hearst 1994), Hearst asked subjects to segment 13 texts of 1800–2500 words, but doesn't report agreement levels. We present results of an empirical study of a large corpus of spontaneous oral narratives, with a large number of potential boundaries per narrative. Subjects were asked to segment transcripts using an informal notion of speaker intention. We found agreement ranging from 82%–92%, with very high levels of statistical significance (from $p = .114 \times 10^{-6}$ to $p \le .6 \times 10^{-9}$).

One of the goals of such empirical work is to use the results to correlate linguistic cues with discourse structure. By asking subjects to segment discourse using a non-linguistic criterion, the correlation of linguistic devices with independently derived segments can be investigated. Grosz and Hirschberg (Grosz and Hirschberg 1992, Hirschberg and Grosz 1992) derived a discourse structure for each text in their study, by incorporating the structural features agreed upon by all of their subjects. They then used statistical measures to characterize these discourse structures in terms of acoustic-prosodic features. Morris and Hirst (Morris & Hirst 1991) structured a set of magazine texts using the theory of Grosz and Sidner (1986). They developed a lexical cohesion algorithm that used the information in a thesaurus to segment text, then qualitatively compared their segmentations with the results. Hearst (1993a) derived a discourse structure for each text in her study, by incorporating the boundaries agreed upon by the majority of her subjects. Hearst developed a lexical algorithm based on information retrieval measurements to segment text, then qualitatively compared the results with the structures derived from her subjects, as well as with those produced by Morris and Hirst. In more recent work, Hearst (1994) presents two implemented segmentation algorithms based on term repetition. The boundaries produced by the algorithms are compared to the boundaries marked by at least 3 of 7 subjects, using   information retrieval metrics. Iwanska (1993) compares her segmentations of factual reports with segmentations produced using syntactic, semantic, and pragmatic information. We derive segmentations from our empirical data based on the statistical significance of the agreement among subjects, or *boundary strength*. We develop three segmentation algorithms, based on results in the discourse literature. We use measures from information retrieval to quantify and evaluate the correlation between the segmentations produced by our algorithms and those derived from our subjects.

3. Intention-Based Segmentation

3.1 Design of Study

If segments are taken to be a definitional construct, i.e., if by definition certain lexical items or other constructions are presumed to indicate the onset or closure of a segment, then the theorist should be at pains to motivate such definitions, e.g., through functional analysis of discourse corpora. But if segments are taken to be an independent construct that correlate directly with the focus of attention (cf. Grosz and Sidner 1986) or the speaker's specific rhetorical purposes (cf. Mann et al. 1992), the definition of segment must be formulated independently of the distribution of linguistic devices.   Only then does the hypothesis that the distribution of certain linguistic devices correlate with discourse segments have substantive content. We take this second tack, using an empirically derived database of segments.

First, a criterion for discourse segmentation must be elucidated. Unfortunately, the correspondence between discourse segments and more abstract units of meaning is poorly understood (cf. Moore and Pollack 1992). A number of alternative proposals have been presented which directly or indirectly relate segments to intentions (Grosz and Sidner 1986), RST relations (Mann et al. 1992) or other semantic relations (Polanyi 1988, Hobbs 1979). In our study, we ask subjects to segment discourse using communicative intention as the segmentation criterion, as in (Grosz and Sidner 1986). Most importantly, by asking subjects to segment discourse using a non-linguistic criterion, the correlation of linguistic devices with *independently* derived segments can be investigated.

A subject pool must then be selected. Discourse segmentation has typically been performed by researchers, rather than obtained experimentally by empirical methods. We use "naive" subjects, as opposed to researchers trained in the area of computational discourse. In particular, the subjects are introductory psychology students at the University of Connecticut and volunteers solicited both from electronic bulletin boards at Columbia University and by the authors. Our decision to use naive subjects provides us with a large pool of subjects, and also helps ensure that subjects are not performing segmentation using the linguistic theories that we hope to independently validate.

The type of segmentation data to be collected must be determined. To simplify data collection, we asked subjects to perform a linear segmentation, as described below. We did not ask subjects to identify the type of hierarchical relations among segments illustrated in Figure 1. In a pilot study we conducted, subjects found it difficult and time-consuming to identify non-sequential relations. Given that the average length of our narratives is 700 words, this is consistent with previous findings (Rotondo 1984) that non-linear segmentation is impractical for naive subjects in discourses longer than 200 words. In addition, issues of intersegment relatedness (e.g., whether there is a standard set of relations that exist between segments) are still poorly understood.

A linear segmentation results from dividing a transcript of a narrative into sequential segments. Our corpus consists of 20 narrative monologues about a movie, taken from (Chafe 1980) ($N \approx 14\,000$ words), each segmented by 7 subjects.[3] Subjects were instructed to identify each point in a narrative where the speaker had completed one communicative task, and begun a new one. These points were thus the boundaries of segments. Subjects were also instructed to identify the speaker's intention associated with each segment. Intention was explained in common sense terms and by example, e.g.:

...

You should think of each movie narration as resulting from many decisions made by the speaker about what to do next. You will be asked to evaluate what the speaker was doing at each point...Read through the transcript and draw a horizontal line across the page between complete text lines (utterances) where you think the speaker started doing something new.

...

In the wide left hand margin, say in abbreviated form what the speaker is doing.

...

[Here] is an example of how to proceed. You are free to use any criteria in deciding what the narrator of your transcript is *doing*.

---

[3] Except in rare cases, no subject segmented more than one narrative.

speaker recommends movie	Well it's really a great movie,
	really beautiful scenery.
	You should see it,
	I recommend it,
	I really do.
	The first part of the movie just sets up...

Figure 2. Excerpt from Instructions

Finally, to simplify data analysis, we restricted subjects to placing boundaries between the prosodic phrases identified in (Chafe 1980). In principle, this makes it more likely that subjects will agree on a given boundary than if subjects were completely unrestricted. However, previous studies have shown that the basic unit subjects use in similar tasks corresponds roughly to a breath group, prosodic phrase, clause, or intonation unit (Chafe 1980, Rotondo 1984, Hirschberg and Grosz 1992, Ono and Thompson, Chapter 3). Thus one effect of using smaller units would have been to artificially lower the probability that two subjects would agree on the exact same boundary site, and conversely, to artificially inflate the significance when subjects agree.

Figure 3 shows the boundary responses of subjects at each potential boundary site for a portion of the excerpt from Figure 1.[4] The identity of each subject is indicated by a letter of the alphabet. (The intentions associated with each subject's segmentations are illustrated later). Line spaces between prosodic phrases represent potential boundary sites. Note that a majority of subjects agreed on only 2 of the 11 possible boundary sites: after 3.3 (n = 6) and after 8.4 (n = 7). (The symbols NP, CUE and PAUSE are explained below.)

Agreement among subjects was far from perfect, as shown by the presence here of 4 boundary sites identified by only 1 or 2 subjects. However, as we describe in subsequent sections, the probabilities of the observed patterns of agreement are extremely low, thus extremely statistically significant.

---

[4] The transcripts presented to subjects did not contain line numbering or pause information (pauses indicated here by bracketed numbers.)

3.3	[.35+ [.35] a-nd] he- u-h [.3] puts his pears into the basket.	
	—6 SUBJECTS (a, b, c, d, e, f)—	NP, PAUSE
4.1	[1.0 [.5] U-h] a number of people are going by,	
		CUE, PAUSE
4.2	[.35+ and [.35]] one is [1.15 um] /you know/ I don't know,	
4.3	I can't remember the first...the first person that goes by.	
	—1 SUBJECT (d)—	PAUSE
5.1	[.3] Oh.	
	—1 SUBJECT (e)—	NP
6.1	A u-m.. a man with a goat [.2] comes by.	
	—2 SUBJECTS (d, e)—	NP, PAUSE
7.1	[.25] It see it seems to be a busy place.	
		PAUSE
8.1	[.1] You know,	
8.2	fairly busy,	
	—1 SUBJECT (c)—	
8.3	it's out in the country,	PAUSE
8.4	[.4] maybe in u-m [.8] u-h the valley or something.	
	—7 SUBJECTS (a, b, c, d, e, f, g)—	NP,CUE, PAUSE
9.1	[2.95 [.9] A-nd um [.25] [.35]] he goes up the ladder,	

Figure 3. Excerpt from 9, with Boundaries

3.2 Reliability Issues

Once data is collected from an empirical study, how can and should it be evaluated? Many data evaluation techniques and methodologies already exist outside the area of computational discourse. We draw on this body of knowledge to provide us with both a vocabulary for defining the issues, as well as with a set of analytic techniques.

We evaluate the segmentation data using *reliability*, as opposed to a weaker notion of *agreement* or a stronger notion of *validity*. Krippendorff's (1980) introduction to content analysis nicely illustrates the meanings of and differences among these terms:

To test *reliability*, some duplication of efforts is essential. A reliable procedure should yield the same results from the same set of phenomena regardless of the circumstances of application. To test *validity*, on the other hand, the results of a procedure must match with what is known to be "true" or assumed to be already

· valid. It follows that, whereas reliability assures that the analytical results represent something real, validity assures that the analytical results represent what they claim to represent....*Reliability data* require that at least two coders independently describe a possibly large set of recording units in terms of a common data language....If agreement among coders is perfect for all units, then reliability is assured. If the agreement among coders is not better than chance...reliability is absent.

In other words, an agreement figure in isolation is often deceptive, since it does not indicate how the figure compares with the agreement figure that would occur by chance. We report the level of agreement in our data, but ultimately use a reliability measure based on comparing the observed distribution relative to a chance model. The validity of the subjects' segmentations is currently beyond our grasp, as the "true" segmentation of a discourse is not independently known. We later refer to the statistically significant boundaries as empirically validated, but by this we mean only that we use them as a target against which to validate how closely algorithm performance can approximate human performance.

3.3 Statistical Results

Since it is less informative than a reliability measure, we omit detailed discussion of percent agreement among our human subjects (but cf. Passonneau and Litman 1993). However, previous work has used percent agreement as a summary statistic to report consistency across subjects performing various types of discourse segmentation (cf. Section 2). For purposes of comparison, we give a brief discussion of our percent agreement results here.

As defined in (Gale et al. 1992), *percent agreement* is the ratio of observed agreements with the majority opinion to possible agreements with the majority opinion. In our empirical study, agreement among from 4 to 7 subjects on whether or not there is a segment boundary between two adjacent prosodic phrases constitutes a *majority opinion*. Given a transcript of length n prosodic phrases, there are n − 1 possible boundaries. Thus the total *possible agreements* with the majority corresponds to the number of subjects times n − 1. Total *observed agreements* equals the number of times that subjects' boundary decisions agree with the majority opinion.

Table 1 presents the data from Figure 3 as a matrix where the 11 columns represent possible boundary slots and the 7 rows represent the subjects' responses. Column totals of 3 or less represent a majority opinion that the slot is a non-boundary; conversely, totals of 4 or more represent a majority opinion that the slot is a boundary. There are $11 \times 7 = 77$ possible agreements with the majority opinion, and 71 observed agreements. Thus, percent agreement for the excerpt as a whole is $^{71}/_{77}$, or 92%. Percent agreement over our corpus averaged 89% ($\sigma = .0006$; max = 92%; min = 82%). The low standard deviation, or spread around the average, also shows that subjects are consistent with one another.

Table 1. Matrix Representation of Boundary Data, Excerpt from 9

Subject	Potential Boundary Slots										
	3.3, 4.1	4.1, 4.2	4.2, 4.3	4.3, 5.1	5.1, 6.1	6.1, 7.1	7.1, 8.1	8.1, 8.2	8.2, 8.3	8.3, 8.4	8.4, 9.1
a	1										1
b	1										1
c	1								1		1
d	1			1		1					1
e	1				1	1					1
f											1
g	1										1
Total	6	0	0	1	1	2	0	0	1	0	7

Non-boundaries, with an average percent agreement of 91% (σ = .0006; max = 95%; min = 84%), showed greater agreement among subjects than boundaries, where average percent agreement was 73% (σ = .003; max = 80%; min = 60%). This partly reflects the fact that non-boundaries greatly outnumber boundaries, an average of 89 versus 11 majority opinions per transcript. Because the average narrative length is 100 prosodic phrases and boundaries are relatively infrequent (average boundary frequency = 15%), percent agreement among 7 subjects is largely determined by percent agreement on non-boundaries. Thus, total percent agreement could be very high, even if subjects did not agree on any boundaries. But percent agreement on boundaries alone is also high. Furthermore, as we show next, the reliability is extremely high.

There are various ways of arriving at a reliability figure. For example, a reliability statistic can quantify how close a set of observations is to a predicted distribution, or how far it is from a random one. Krippendorff (1980) presents an agreement coefficient that indicates the extent to which an observed distribution such as that of Table 1 resembles a maximum rather than a chance agreement distribution (where chance agreement is estimated from the coders' distributions). His method also allows subjects to make more than binary distinctions in their decisions. In addition, there are various ways to model randomness of a data set, depending on the dimensions that are allowed to vary freely. We use Cochran's Q (Cochran 1950) to evaluate the reliability of our segmentation data.[5] Cochran's Q is a summary statistic that compares an observed matrix such as Table 1 to a random model given by allowing the locations of a subject's boundaries to vary, but restricting the total number of boundaries assigned by any subject to be equal to the observed total. In other words, column totals can vary randomly, but row totals are fixed.

---

[5] We thank Julia Hirschberg for suggesting this test.

We represented our segmentation data as a set of i × j matrices, such as Table 1 with height i = 7 subjects and width j = n − 1 prosodic phrases. Thus a row total corresponds to the total number of boundaries assigned by subject i. In Table 1 (j = 11), subject d assigned 4 boundaries. The probability of a 1 in any of the j cells of the row is thus $^4/_{11}$, with $(^{11} _4)$ ways for the 4 boundaries to be distributed. Taking this into account for each row, Cochran's test evaluates the null hypothesis that the number of 1s in a column, here the total number of subjects assigning a boundary at the jth site, is randomly distributed. Where the row totals are u_i, the column totals are Tj, and the average column total is Γ, the statistic is given by:

$$Q = \frac{j\,(j-1)\,\Sigma\,(T_j - \Gamma)^2}{\chi\,(\Sigma\,u_i) - (\Sigma\,u_i^2)}$$

Q approximates the χ^2 distribution with j − 1 degrees of freedom (Cochran 1950). Our results indicate that the agreement among subjects is extremely highly significant. That is, the number of 0s or 1s in certain columns is much greater than would be expected by chance. For the 20 narratives, the probabilities of the observed distributions range from $p = .114 \times 10^{-6}$ to $p \le .6 \times 10^{-9}$.

The potential boundary sites can be sorted into two classes, boundaries versus non-boundaries, depending on how the majority of subjects responded. Partitioning Q by the 7 values of j shows that Q_j is significant for each distinct T_j ≥ 4 across all narratives (cf. Passonneau and Litman 1993). For T_j = 4, $.0002 \le p$ $\le .30 \times 10^{-8}$; probabilities become more significant for higher levels of T_j, and the converse. At T_j = 3, p is sometimes above our significance level of .01, depending on the narrative. Thus we use T_j = 4 as the threshhold for distinguishing empirically validated boundaries from non-boundaries.

3.4 Fuzzy Boundaries

Despite the statistically significant agreement among subjects in identifying common segment boundaries, there is a lot of variability in the number and precise locations of segments proposed by each subject. Figure 4 illustrates a case in point. It shows 7 prosodic phrases from narrative 12. Seven subjects agreed on boundary A, located at phrase (15.1,16.1). Two subjects located boundary B at (16.4,17.1), and three different subjects located it at (17.1,18.1). Thus a majority agree that boundary B occurs somewhere in the same vicinity, although no location meets the threshhold (T_j ≥ 4) established above.

In addition, as discussed in Section 5, the four or five phrases between A and B describe a coherent sub-event of a well-defined episode of the Pear narratives. This constitutes conceptual motivation for classifying them as a coherent segment despite the imprecision in the location of boundary B. We are still developing a reliable method for integrating fuzzy boundary locations into our analysis (cf.

Section 3.5). For present purposes, however, we quantify correlation of linguistic cues using data like boundary A, where a majority of subjects agreed on the precise location. In the next subsection, we describe the information retrieval metrics we use to evaluate the performance of an individual subject or algorithm against the consensus. Then in the following subsection, we present data quantifying the range of variation among subjects. This provides a baseline for interpreting the results for the algorithms.

15.1   [3.1 [7.5] A-nd [.45] then [1.15 tsk..]] he [.45] the goat [.25] and the goatman...disappeared.

A   —7 SUBJECTS (a, b, c, d, e, f, g)—

16.1   [.55] A young boy on a bicycle,
16.2   [.45] that was much too big for him,
16.3   [1.35] rode [.45] thee.. from the [.2] direction in which the goat [.25] person had come,
16.4   [.8] towards the man picking the [.75] the pearpicker [3.0 [1.45.. tsk..]] a-nd [1.25]] stops.

B   —2 SUBJECTS (a, c)—

17.1   [1.3] Beside the baskets.

B   —3 SUBJECTS (d, e, g)—

18.1   [.5] While the man was upsta he had gone up the tree.

Figure 4. Indeterminacy of Boundary Location

3.5 Use Of Information Retrieval Metrics

Information retrieval (IR) metrics are designed to quantify, e.g., what proportion of documents retrieved by a given query are *correct*, what proportion of *desired* documents are retrieved, and so on. We use IR metrics to quantify what proportion of statistically validated boundaries are recognized, and so on. We make two caveats about the use of IR metrics. First, IR metrics are overly conservative, given the binary classification of ideal and actual performance into recognition of boundaries versus non-boundaries. As noted above, such a strict classification fails to address near misses like boundary B in Figure 4. In future work we will remedy this by using sums of boundary weights within a moving window of 3 or more boundary slots, or by using string matching techniques. Second, interpreting the values of the IR metrics requires some notion of how hard the task is. In the next subsection, we establish a baseline for comparison by quantifying the performance of human subjects using the IR metrics.

Figure 5 schematically presents the distributions used to compute four IR metrics. The column marginals, or sums, (i.e., a + c and b + d) represent the classification of possible boundary locations (pairs of sequential prosodic phrases) into empirically validated boundaries and non-boundaries. The row marginals (i.e., a+b and c+d) represent the hypotheses of an individual subject or algorithm for a given narrative regarding boundaries and non-boundaries. Each cell represents an intersection of an ideal classification with a hypothesized one, thus 'a' represents the set of correctly hypothesized boundaries, 'b' represents the set of incorrectly hypothesized boundaries, and so on. The formulas for the four IR metrics we use to summarize such a table are also given in Figure 5. Ideally, the number of incorrect hypotheses for boundaries (c) and non-boundaries (b) both equal 0, giving a value of 1 for both recall and precision, and a value of 0 for both fallout and error rate. We will see in the next section that average human performance is far from ideal.

Test Responses	Empirically Validated Responses	
	Boundary	Non-Boundary
Boundary	a	b
Non-Boundary	c	d

Recall = a / (a + c) ratio of correctly hypothesized boundaries to empirically derived boundaries	Fallout = b / (b + d) ratio of incorrectly hypothesized boundaries to empirically derived non-boundaries
Precision = a / (a + b) ratio of correctly hypothesized boundaries to hypothesized boundaries	Error = (b + c) / (a + b + c + d) ratio of incorrect hypotheses to the table total

Figure 5. Information Retrieval Metrics

3.6 Baseline Human Performance

Ultimately, we quantify the correlation of linguistic devices with segment boundaries using IR metrics. However, interpreting the results depends on first understanding the inherent variability in the human subjects' data.

Given a particular narrative, we take boundaries agreed on by a majority of subjects to be empirically validated. Since the seven subjects providing data on any one narrative all assign boundaries at a different rate, the performance of any one subject is likely to deviate from the consensus. Table 2 gives an overall picture of the human subjects' performance. Row 1 repeats ideal scores for the 4 metrics (cf. Figure 5), for convenient reference. The average performance across all subjects on all narratives, given in row 2, shows in particular that average precision is relatively poor, with roughly only half the hypothesized boundaries being 'correct'. This results from the wide variation in the rate at which subjects assign boundaries. The average frequency of boundary assignment is 15%, but the standard deviation (σ = .07) reflects the large spread around the average (max = .41; min = .03). When a subjects' rate is higher than average, the value of 'b' (cf. Figure 5) increases and precision decreases.

Table 2 also shows the best and worst values for each metric achieved by any subject (rows 4 and 6), and the best and worst overall performance for a given subject (rows 5 and 7). To rate overall performance for a particular subject, we compute a sum of deviations (column 6): the sum of differences between the observed and ideal values for each metric. In the worst possible performance, recall and precision equal 0 instead of 1, and fallout and error equal 1 instead of 0, thus the summed deviation would equal 4. Note that the values for the best metrics are close to ideal for recall (1), fallout (.01) and error (.04). The subject with the best overall score also was the single subject with the best precision (.86), and notably achieved both high precision and recall. This subject assigned 7 boundaries, 6 of which were validated boundaries, in a relatively short narrative that had a total of 7 validated boundaries. Note that although this subject had only 1 type 'b' error and one type 'c' error, recall and precision are not higher because there were relatively few boundaries.

Table 2. Individual Human Performance

	Recall	Precision	Fallout	Error	Deviation
Ideal	1.00	1.00	.00	.00	.00
Average	.74	.55	.09	.11	.90
s	.20	.17	.06	.05	.34
Best Metric	1.00	.86	.01	.04	.19
Best Subject (narr 3)	.86	.86	.02	.04	.34
Worst Metric	.20	.09	.35	.33	2.39
Worst Subject (narr 6)	.20	.09	.13	.17	2.01

A single subject happened to have both the worst fallout (.35) and the worst error rate (.33), but had a relatively high recall (.81) and moderately bad precision (.26; i.e., roughly midway between the worst of .09 and the average of .55). This reflects this subject's very high rate of assigning boundaries (41.0% compared with the average of 22.7%), leading to higher than average errors of both types 'b' and 'c'. The subject with the worst recall (.2) and precision (.09) had only moderately bad fallout (.13) and error rate (.17) because, having a boundary rate equal to the average, this subject's errors of type 'b' and 'c' were still low relative to 'd' despite being high relative to 'a'. Thus interpreting Table 2 requires an understanding of the interdependencies among the IR metrics and the typically low rate of assigning boundaries.

4. Correlation of Boundaries and Linguistic Cues

In the preceding section, we demonstrated that the reliability of humans' ability to recognize segment boundaries is highly significant, despite the open-endedness of the task presented to the subjects, despite what is mostly likely an overly rigid notion of *segment boundary location*, and despite the variability of human performance. Our next goal is to determine whether segments reflect processes or structures that would be useful for generating or understanding extended discourse, regardless of their role in on-line human processing. We tackle this goal in two ways. First, in this section, we quantify the degree to which linguistic features correlate with segment boundaries, thus providing automatic means for recognizing or generating segment boundary location.

4.1 Three Algorithms

In principle, the process of determining whether the empirically validated segment boundaries correlate with linguistic devices would require a complex search through a large space of possibilities, depending on what set of linguistic devices one examines, and what features are used to recognize and classify them. Our interest is primarily in providing an initial evaluation of existing hypotheses, rather than in developing a method to search blindly through the space of possibilities. Thus we have begun by looking at three devices whose distribution or surface form has frequently been hypothesized to be conditioned by segmental structure: referential NPs, cue words and pauses.

Referential Noun Phrases

Distributional analyses of discourse anaphoric NPs in various kinds of monologic, conversational or textual discourse has generated specific hypotheses about how surface form correlates directly or indirectly with segmental structure (cf., e.g., Grosz 1977, Sidner 1979, Passonneau 1985, Grosz and Sidner 1986, Reichman 1985, Levy 1984, Passonneau 1992, Passonneau 1994). We use a segmentation algorithm based on features of the surface form of referential NPs developed out of our previous work on formulating specific discourse constraints. For further details on the motivation for the algorithm design, see (Passonneau and Litman 1993) and (Passonneau 1993).

The input to the NP algorithm consists of lists of 4-tuples representing all the referential NPs in each narrative, where each tuple is <LOC, NP, INDEX, INFER>. LOC stands for location, and is represented in terms of the % sequential position of the NP, and the sequential position of the prosodic phrases and syntactic clauses in which the NP occurs.[6] NP stands for the surface form of the noun phrase, permitting identification of definite NPs and their classification into phrasal versus pronominal NPs. INDEX stands for the discourse entity evoked by each NP (cf. Karttunen 1976, Webber 1978, Heim 1983, Kamp 1981). If an NP evokes evokes an entity that is already in the discourse model, it is assigned the same index. Finally, INFER is a set of inferential relations between the denotation[7] of the NP and other objects in the discourse context. It has long been noticed that discourse anaphoric NPs are also licensed given a strong inferential link to an existing entity (cf. Grosz 1977, Prince 1981b, Carlson 1977, Hawkins 1978, Hobbs 1979). The output is a set of boundaries B, represented as ordered pairs of adjacent clauses: (C_n, C_{n+1}). This output is then mapped to pairs of adjacent prosodic phrases.[8]

Essentially, the NP algorithm has three tests for determining whether the next pair of adjacent clauses (C_i, C_{i+1}) represents a boundary. Any NP can provide a direct link to the preceding clause through coreference or an inferential relation; a definite pronoun can provide a link to the current segment as a whole. Specifically, a clause pair (C_i, C_{i+1}) is not a boundary under one of the following conditions:

- coreference: the same INDEX is used for an N-tuple in C_{i+1} and an N-tuple in C_i
- an inferential link: an INFER relation for an NP-tuple in C_{i+1} links an INDEX in C_{i+1} to an INDEX in C_i
- definite pronominal reference to an entity in the current segment: a third person definite pronoun in C_{i+1} receives an INDEX that is also used for some NP-tuple from any prior clause, up to the last boundary that was assigned.

If all 3 tests fail, C_{i+1} is determined to begin a new segment, and (C_i, C_{i+1}) is added to the current set of boundaries.

The excerpt in Figure 3 shows two boundaries assigned by the NP algorithm. Boundary (3.3,4.1) is assigned because the sole (non-pronominal) NP in 4.1, *a number of people*, evokes an entity that is entirely new to the discourse, and this entity cannot be inferred from any entity mentioned in 3.3. Thus all 3 tests fail. The next boundary that is assigned is (8.4,9.1). The entity evoked by the full NP *the ladder* is not evoked in 8.4, and it is not inferentially linked to any of the

---

[6] The syntactic clauses consist roughly of tensed clauses that are neither verb arguments nor restrictive relative clauses.

[7] In theory, relevant inferential relations may involve either the discourse entity evoked by an NP or the intended referent (cf. Passonneau 1994), but for present purposes we conflate the entity and referent indices.

[8] Because potential boundary locations assigned by the NP algorithm (between adjacent clauses) can differ from the potential boundary locations for the human study (between adjacent prosodic phrases), the data is first normalized, as described in (Passonneau and Litman 1993). Normalization eliminates a total of 5 boundaries in 3 of the narratives (out of 213 in all 10).

entities in 8.4, thus failing the first two tests. The third person pronoun *he* evokes an entity, the farmer, that was last mentioned in 3.3, and 1 NP boundary has been assigned since then, thus failing the third test.

Cue Words

Cue words (e.g., *now*) are words that are sometimes used to explicitly signal the structure of a discourse. Our segmentation algorithm based on cue words uses a simplification of one of the features shown by Hirschberg and Litman (1993) to identify discourse usages of cue words. Hirschberg and Litman examine a large set of cue words proposed in the literature and show that certain prosodic and structural features, including a position of first in prosodic phrase, are highly correlated with the discourse uses of these words. The input to our cue word algorithm is a sequential list of the prosodic phrases constituting a given narrative. The output is a set of boundaries B, represented as ordered pairs of adjacent phrases (P_n, P_{n+1}), such that the first item in P_{n+1} is a member of the set of cue words summarized in Hirschberg and Litman (1993). That is, if a cue word occurs at the beginning of a prosodic phrase, the usage is assumed to be discourse and thus the phrase is taken to be the beginning of a new segment. Figure 3 shows 2 boundaries (CUE) assigned by the algorithm, both due to *and*. Because *and* occurs frequently in phrase initial position, the cue algorithm assigns boundaries at a high rate. As noted below, this results in low precision.

Pauses

Grosz and Hirschberg (Grosz and Hirschberg 1992, Hirschberg and Grosz 1992) found that in a corpus of recordings of Associated Press news texts, phrases beginning discourse segments are correlated with duration of preceding pauses, while phrases ending discourse segments are correlated with subsequent pauses. We use a simplification of these results in our algorithm for identifying boundaries in our corpus using pauses. The input to our pause segmentation algorithm is a sequential list of all prosodic phrases constituting a given narrative, with pauses (and their durations) noted. The output is a set of boundaries B, represented as ordered pairs of adjacent phrases (P_n, P_{n+1}), such that there is a pause between P_n and P_{n+1}. Unlike Grosz and Hirschberg, we do not currently take phrase duration into account. In addition, since our segmentation task is not hierarchical, we do not note whether phrases begin, end, suspend, or resume segments. Figure 3 shows seven boundaries (PAUSE) assigned by the algorithm.

4.2 Performance of Algorithms

In this section we discuss the performance of the three algorithms first to evaluate their success at locating segment boundaries, but then to draw generalizations regarding the nature of segment boundaries. Unsurprisingly, the performance of the three algorithms as quantified by the IR metrics varies with the amount of knowledge they exploit. More enlightening than comparing the algorithms with one another is to compare them with human performance. We first present and discuss the performance of each algorithm averaged across the narratives as compared with the average of human subjects. The results suggest that levels approximating average human performance could eventually be achieved. However, the key lesson we draw from analyses of variation in the data is that no single strategy for identifying segments would be sufficient. Individual discourses vary significantly in how reliably humans or algorithms can segment them (variation across speakers). Also, there may be inherent limitations to algorithms that rely on a consistent strategy, because boundaries within the same discourse may be signalled by different means (variation within speakers).

Table 3 shows the performance of the three algorithms (NP, CUE, PAUSE) using the 4 IR metrics, along with the summed deviation. Again, the summed deviation ($0 \leq$ deviation ≤ 4) is the sum of the deviations of each metric from ideal performance. For convenient comparison, the first row in Table 3 repeats the human subjects' average performance. The remainder of Table 3 consists of 3 subtables, with 4 rows per algorithm. The first row of each subtable gives average values across the 10 narratives, and the second gives the standard deviation.[9] We use the summed deviations to identify the individual narratives on which each algorithm performs best and worst, shown in the 3rd and 4th rows of the subtables. NP exploits more knowledge than CUE and PAUSE, reacting to one of 3 conditions instead of a single one. This is reflected in the performance differences, with NP performing less well than humans but better than CUE and PAUSE. The average summed deviation for NP (1.53) is almost twice as bad as for humans (.79), but is roughly 75% that for PAUSE (1.93) and CUE (2.16). Recall and precision show the biggest difference between human performance and NP; they are also more sensitive to small changes in performance because values of 'a', 'b', and 'c' tend to be much smaller than values of 'd'. NP recall is .50 compared with .74 for humans; NP precision is .30 compared with .55 for humans. NP error and fallout are closer to human performance (.15 vs. .09; .19 vs. .11). PAUSE (.92) has much higher recall than humans, but PAUSE and CUE have very low precision (.15 and .18) and very high fallout and error rate (ranging from .49 to .54).

---

[9] The figures for NP differ from those in (Passonneau and Litman 1993), where we inadvertently gave the metrics for boundaries defined by a threshold of agreement among at least 5 instead of 4 subjects. Another difference is that formerly the algorithm was applied by hand; here, the coding of NP-tuples is still only partly automated, but application of the algorithm is fully automatic.

Table 3. Evaluation for $T_j \geq 4$

	Recall	Precision	Fallout	Error	Deviation
Avg. Human	.74	.55	.09	.11	.79
Avg. NP	.50	.30	.15	.19	1.53
s	.17	.10	.06	.06	.29
Best (narr 3)	.71	.50	.10	.13	1.02
Worst (narr 2)	.44	.20	.27	.31	1.94
Avg. Cue	.72	.15	.53	.50	2.16
s	.17	.06	.07	.07	.27
Best (narr 3)	.86	.22	.44	.40	1.76
Worst (narr 2)	.40	.04	.60	.60	2.76
Avg. Pause	.92	.18	.54	.49	1.93
s	.09	.05	.06	.06	.17
Best (narr 3)	.94	.23	.48	.42	1.73
Worst (narr 2)	.73	.16	.57	.53	2.21

A second dimension to consider in observing that NP performance is closer to human performance than PAUSE and CUE is that the NP algorithm and human consensus result in relatively fewer boundaries. On average, subjects assigned boundaries 15% of the time (σ=.07). Given the average narrative length of 102 boundary slots, this makes each segment an average of 15 phrases long. NP assigns boundaries using narrow criteria reflecting an assumption that in the default case, each subsequent clause of a narrative continues the current segment rather than starting a new one. It assigns boundaries at a rate of 19%. In contrast, CUE and PAUSE rely on features that occur frequently, and consequently assign boundaries relatively often.

The difference in performance between the NP algorithm versus CUE and PAUSE suggests that by exploiting even more knowledge performance might be further improved. In fact, we are optimistic that all three algorithms can be improved by discriminating among duration of pauses (cf. Grosz and Hirschberg 1992), types of cue words (e.g., signalling pops Grosz and Sidner 1986), and by adding further features to differentiate uses of referential NPs (e.g., grammatical role information, cf. Passonneau 1992).

As another illustration of how increased knowledge can enhance performance, Table 4 presents results for various composite algorithms that locate a boundary wherever two or more of NP, PAUSE and CUE locate a boundary. Precision is the most likely metric to be improved. For a composite algorithm, recall cannot be increased: if neither NP, PAUSE nor CUE found a boundary, then no combination of them can. As noted, recall of combined algorithms is limited by the lowest recall of the individual algorithms: the average recall of .49 for NP/(CUE or PAUSE) (row 5) is only somewhat lower than the average recall of .50 for NP alone and the average recall of .69 for the PAUSE/CUE algorithm (row 3) is somewhat lower than that of .72 for CUE alone. However, the composite algorithms use narrower criteria for boundaries, which should reduce the number of false positives. The precision of the combined algorithms is indeed higher than any of the algorithms alone. NP/PAUSE has the best combined algorithm performance as measured by the summed deviation. We believe it should be possible to algorithmically achieve the performance levels of the average human subject by some combination of additional knowledge and judicial use of multiple knowledge sources. However, the differences across and within narratives indicate inherent limitations to this approach.

Table 4. Combined Algorithms

	Recall	Precision	Fallout	Error	Deviation
NP/Pause	.47	.42	.08	.13	1.42
NP/Cue	.36	.34	.09	.15	1.59
Pause/Cue	.69	.29	.29	.29	1.66
NP/Cue/Pause	.34	.47	.05	.12	1.43
NP/(Cue or Pause)	.49	.34	.12	.17	1.52

Standard deviations (σ) of the various metrics for human performance are included in Table 2 and for algorithm performance in Table 3. Overall, the standard deviations for recall and precision are higher than for fallout and error. This reflects the high proportion of non-boundaries in the data, i.e., values of 'd' are always relatively large, minimizing the effect of errors relative to 'd'. The largest spread occurs among humans: for recall, σ=.20, and for precision, σ=.17. There is wide difference among subjects, even on the same narrative.   For example, for narrative 3, 2 subjects achieved maximum recall of 1 whereas recall of a third subject was equal to the observed minimum of .2. Differences among subjects on the same narrative presumably result in part from different levels of attention and ability on the part of the subjects. However, since subjects bring different goals and different kinds of knowledge to the segmentation task, the variation must also partly reflect different interpretations of the same discourse (cf. Section 2). As we note in the conclusion, this is an issue that should be addressed in designing natural language understanding systems.

As noted earlier, the performance differences in the algorithms primarily reflects differences in the amount of knowledge they use. But performance differences across narratives may also reflect differences in the narrative strategies used by the speakers. Levy (1984), for example, identifies several distinct speaker strategies for reidentifying major and minor characters based on an analysis of narrative monologues about the television dramatization of *Brideshead Revisited*. On the one hand, some speakers seem to produce narratives that because of greater consistency in speaker strategy, or other factors to be determined, are segmented more reliably by more than one algorithm. Narrative 3, for example, yields the best scores for NP and CUE as well as the best scoring human. On the other hand, some speakers may rely more exclusively on a strategy that affects one type of linguistic feature, making other surface features irrelevant. This might explain why narrative 2 shows the worst performance for NP but the best performance for PAUSE.

To summarize our interim conclusions regarding the correlation of linguistic cues with segment boundaries, our results show that it may be possible to replicate average human performance, but that doing so would require multiple sources of knowledge. For example, the NP algorithm looks at multiple features of NPs, and its best performance (recall=.71; precision=.50; fallout=.10; error rate=.13) is close to the average of humans performance (recall=.74; precision=.55; fallout=.09; error rate=.11). Combining NP and PAUSE raises average precision by a third and causes small improvements in fallout and error rate without significantly degrading recall. However, a close look at different aspects of variation in the subjects' reliability at identifying boundaries, and of the spread in performance on individual narratives suggests that in addition to using multiple knowledge sources, automated methods for recognizing segments through surface features would ultimately need to adapt dynamically to different strategies within and across speakers.

5. From Boundaries to Segments

The previous section asked whether specific locations for segment boundaries can reliably be identified by humans or automated methods. In this section, we address the question of how the results on segment boundaries pertain to coherence. First, we illustrate the degree to which subjects agreed on the intentions they assigned to segments by examining one segment in detail. Then we examine two multi-segment phenomenon. In one case, we take a semantically coherent multi-segment unit consisting of an episode mentioned in all 10 narratives. We discuss the relation of segment boundaries to segments in this episode, focussing on the issue of fuzzy boundary location. We re-examine one of the algorithms on the episode, using fuzzy boundary location. Finally, we examine the more general phenomenon of discourse pops, illustrated in our first example in Figure 1, where the relation among distinct segments is pragmatically defined. We discuss the nature of the segment boundaries in all the pops involving a definite pronoun in the resumption segment. We also discuss the semantic coherence of the segment that is popped over.

5.1 Discourse Segment Intention

Inspection of subjects' descriptions of speaker intention suggests that subjects not only agree on the boundaries of segments, but also agree on the narrator's intentions for the segment. As yet we have not found a suitable formal methodology for analyzing the reliability of the data regarding intentions. For example, the matrix for representing boundary decisions (recall Table 1) is not suitable for representing segment intentions. In particular, since each subject associates intentions with a different set of segments, it is not possible to have the same set of columns across subjects when specifying intentions. In addition, when deciding whether a potential site is a boundary or not, subjects are restricted to a dichotomous choice. In contrast, subjects are not restricted to choosing intentions from a pre-defined set of values. We informally examine the qualitative agreement among subjects on identification of speaker intention.

12.5   [2.1 [.2] a-nd [1.2]] his [.3] bicycle hits a rock.

—1 SUBJECT (g)—

13.1   Because he$_i$'s looking at the girl.

—1 SUBJECT (f)—

14.1   [.75] ZERO-PRONOUN$_i$ Falls over,

—5 SUBJECTS (a, b, c, d, e)—

14.2   [1.5 [1.35] uh] there's no conversation in this movie.
15.1   [.6] There's sounds,
15.2   you know,
15.3   like the birds and stuff,
15.4   but there... the humans beings in it don't say anything.

—7 SUBJECTS (a, b, c, d, e, f, g)—

16.1   [1.0] He$_i$ falls over,

Figure 6. Excerpt from 6

Consider the segment that can be derived from Figure 6. The segmentation shown contains 2 boundaries proposed by one subject each, 1 boundary proposed by five subjects, and 1 proposed by all seven subjects. The first 3 of these comprise a fuzzy boundary in the sense illustrated above with Figure 4; 7 distinct subjects locate a boundary at or near (14.1,14.2).   This yields a single discourse segmentation for the excerpt spanning from an initial fuzzy boundary at (12.5,13.1,14.1,14.2) to (15.4,16.1). However, the subject who located the initial boundary earlier than the 6 other subjects, 'g' at (12.5,13.1), associates a completely distinct intention with the segment.

Figure 7 presents the intentions attributed to the narrator by the 7 subjects. The 5 subjects who agreed on exactly the same segment boundaries (a–e) all indicate that the speaker's intentions pertain to describing the audio characteristics of the movie. Subject f identifies the segment to include one more phrase (cf. Figure 6), and characterized the segment purpose as *techniques used in the movie*. Thus f identifies essentially the same intentional structure. However, subject g, who began the segment with 13.1, offers an intention that has no relevance to the sound track or cinematic properties of the movie. Subject g thus not only identifies a different segment boundary, but also a different overall purpose.

Subject	Annotation of Narrator's Intention
a	Digression to describe sound track
b	No verbal communication [i.e., speaker describes lack]
c	Describes that it is a movie with only nature sounds
d	Speaker describes sound techniques used in movie
e	Explain that there is no speaking in movie
f	Techniques used in the movie
g	Because he's looking at a girl when the boy falls no one cares about him

Figure 7. Segment spanning 14.2 through 15.4

Presumably, the first 6 subjects depicted in Figure 7 abstract from the fact that the three full clauses in the segment all refer to auditory characteristics, as signaled by the lexical items *conversation* in the first clause, *sounds* in the second, and *say* in the third. Here we have generalized further from the subjects annotations to note that each asserts something about the movie's auditory character. In general, for segments delimited by high agreement boundaries, a single formulation of speaker purpose can be generalized from the data provided by the seven subjects. We believe that such data provides evidence that when asked to, subjects perform the same kinds of abstraction across related utterances described in (Polanyi 1988) and elsewhere.

5.2 The Theft Episode

One of the key events in the narrative is an episode in which a young boy, one of the main characters in the narrative, steals a basket of pears. We use this episode to illustrate the interaction between the sometimes imprecise location of segment boundaries, and segment content. We show that with a fuzzier definition of boundary, the NP algorithm performance improves significantly, particularly on precision. However, variation across narratives remains high, reinforcing our conclusion that maximum performance could not be achieved without relying on multiple types of features.

By abstracting across the descriptions of the theft episode in our 10 narratives in a manner similar to the method described by Polanyi (1988), the 8 actions listed in Figure 8 can be identified. In all but one narrative, the episode consists of a contiguous sequence of prosodic phrases. Earlier, we used Figure 4 to illustrate a fuzzy boundary. Recall that five subjects agreed that a boundary occurred within 2 adjacent boundary slots: (16.4,17.1) or (17.1,18.1). If this were a boundary, it would separate a segment describing the first event listed in Figure 8 from a segment describing event 2 of the episode. Here we define a fuzzy boundary to be a contiguous sequence of inter-phrasal locations PP_i to PP_n such that the total number of distinct subjects identifying PP_i to PP_n is at least 4. By this definition, boundary B in Figure 4 is a fuzzy boundary beginning at (16.4,17.1) (2 subjects) and ending at the next adjacent location (17.1,18.1) (3 subjects). Using fuzzy boundaries adds 3 boundaries out of a total of 46 for all the segments where the theft episode is described.

1. boy arrives on bike
2. boy steals the one full basket of pears
3. boy rides away with basket balanced on bike handlebars
4. boy is distracted by an approaching girl
5. boy loses hat
6. boy's bike hits rock
7. boy falls over
8. pears spill all over the ground

Figure 8. Events in the Theft Episode

Using fuzzy boundaries, the average number of segments spanning the sequence is 3.6. The range is from 1 to 11, so there is considerable variation in how many utterances or segments narrators used in relating the episode. One uniformity is that relatively more of the boundaries that are external to the theft episode, i.e., those that begin or end the whole episode, are identified by 6 or more subjects: 18 of the 22 external boundaries.[10] Conversely, 15 of the 24 episode internal boundaries are identified by at most 4 or 5 subjects. The average number of subjects identifying an external boundary is 6.5, compared with 5.3 for internal boundaries.[11] Thus subjects are more likely to identify a boundary marking the break between this episode and other material than one marking intra-episode breaks.

The NP algorithm is also more likely to find an external boundary than an internal one: it finds 17 of the 22 external boundaries but only 10 of the 24 internal ones. This provides a possible functional correlation for our suggestion in (Passonneau and Litman 1993) that boundaries identified by more subjects are located more reliably by all 3 algorithms; i.e., location of more salient boundaries may correlate with more global discourse intentions.

Table 5 contrasts three performance figures for the NP algorithm: the overall performance figures from Table 3 (row 1), figures for the theft episode alone, but using the same strict notion of boundary location (row 2), and figures for the theft episode using the fuzzy definition of boundary location (row 3). For fuzzy boundaries, we changed the output of the algorithm as follows. Contiguous boundary slots identified by the algorithm were classified as a single boundary;

---

[10] Narrative 1 has more external boundaries because of an interruption within the episode.

[11] Similar differences between external and internal boundaries exist using the strict notion of boundary.

thus whether the algorithm identified one or both locations for B in Figure 4, it counts as one right answer and no wrong answers. This both penalizes and aids the algorithm: it degrades recall because it becomes impossible to find the occasional occurrences of two distinct boundaries in adjacent slots, but it improves precision.

Table 5 shows an improvement in recall (from .5 to .6) and precision (from .3 to .4), using the strict definition of boundary. We believe this is because in all cases but one narrative, the input consists of a continuous sequence of phrases that is by definition more coherent: it comprises a single episode. Within a whole narrative, there are occasional sequences of phrases whose function is more difficult to interpret, and which are therefore harder to segment. Table 5 also shows an event greater improvement when the input is both a single episode, and boundaries are fuzzy. All scores improve, with precision unsurprisingly showing the greatest improvement.[12]

Perfect scores were achieved on the one narrative (3) in which the whole episode is comprised by a single segment. This segment consisted of 14 phrases. As illustrated in Table 5, the worst performance was on narrative 17, which had 5 segments in the episode (with no difference between strict and fuzzy conditions). Lest one assume that the performance degrades as the number of segments increases, performance was very good for another narrative excerpt consisting of 5 segments: for narrative 12, recall was .80, precision was .67, fallout was .05 and error rate was .07 with a summed deviation of .65. In fact, the performance of the algorithm is far better on 8 of the narratives than on 2 and 17. For example, the average summed deviation of these 8 narratives is only .56, compared with .96 for 2 and 1.74 for 17.

Initial experiments suggest that improvements to NP would be easier for the 8 narratives on which it already performs well, than on 2 or 17. One condition that seems to improve performance is to require NP to assign a boundary if a completely new entity is introduced that is not inferentially related to entities in the previous utterance. By adding this feature, recall in the 8 narratives would be nearly perfect.[13] However, the effect on narratives 2 and 17 would be quite different. This feature is associated with 4 of the 5 boundaries missed by the algorithm in narrative 17, but with only 1 of the 6 boundaries missed by the algorithm in narrative 2. Recall also that narrative 2 was the narrative yielding the worst overall score for NP (cf. Section 4.2), with a summed deviation of 1.94. Thus, the narrators in 2 and 17 not only use different patterns of expression from the remaining 8, as illustrated by differences in performance of NP on the theft episode, but also differ from one another.

---

[12] Hearst (1994) also finds a big improvement when she admits fuzzy boundaries.

[13] Since the use of new features is still under development, we do not have actual scores on metrics for an NP algorithm using this feature.

Table 5. NP Performance on Theft Episode

	Recall	Precision	Fallout	Error	Deviation
Whole Narrative	.50	.30	.15	.19	1.53
Strict	.60	.40	.17	.20	1.37
Fuzzy	.69	.74	.04	.09	.70
Best Fuzzy (3)	1.00	1.00	.00	.00	.00
Worst Fuzzy (17)	.17	.33	.06	.18	1.74

In sum, NP performance improves when the input is semantically more coherent, and when boundaries are allowed to have fuzzy locations. However, large differences in performance from one narrative to another still exist.

5.3 Discourse Segment Pops

Although our subjects were not asked to specify structural or semantic relations among segments, here we briefly look at a specific class of the structural discourse relationship *pop*, which we can identify linguistically from our data. In the general case, discourse pops occur when an earlier suspended segment is resumed with a later segment. In this section we restrict our attention to pops where the first complete utterance of the segment contains a third person definite pronoun whose referent is not mentioned in the preceding segment, but rather in a prior segment. The pop in Figure 1 is an example. Such pops often occur after a segment on which there is strong consensus of a semantic discontinuity. Since the resolution of the reference of the pronoun depends on recognizing a relation between the new segment and a previously suspended segment, we raise the question of how such pops are recognized, and whether the semantic relations are directly or indirectly encoded.

In Fig. 6, one of the main characters of the movie (a boy on a bicycle) is referred to in phrases 13.1 and 14.1. The speaker suspends her description of the boy's activities throughout the next segment ([14.2,15.4]), but resumes reference to the boy in the first utterance of the following segment (16.1) using a third person definite pronoun subject (*he*). This illustrates a specific class of discourse pops, in which a segment resumes an earlier *suspended segment*. In particular, this class of *resumption segments* begins with an utterance in which a third person definite pronoun refers to an entity that is not in the focus space (Grosz and Sidner 1986) *intervening segment*. Processing a clause that signals this type of discourse pop involves several tasks. Resolving the pronoun in the initial clause of the resumption segment requires shifting the attentional state (Grosz and Sidner 1986), since the active focus space (corresponding to the intervening segment) does not contain a representation of the referent. This shift depends on recognizing the termination of the intervening segment, and a continuation relation between the resumption and suspended segments, so that the entities in focus in the suspended segment are again in focus for the resumption segment. There are 8 discourse pops of this type in the 10 narratives we tested the algorithms on. These 8 examples exhibit various structural and semantic relations to the presumed discourse model. For example, they contain intervening segments that provide more detail, provide general background, or are digressions.

In 7 cases, the resumption segment begins with a word that can function as a cue word,[14] but in four cases the cue word is *and*, a word whose discourse usage is hard to distinguish and which provides very little semantic information. The cue words in the remaining 3 cases are *so, all right* and *then*, none of which clearly signal attentional change (Grosz and Sidner 1986, Hirschberg and Litman 1993). Non-lexical signals (e.g., *uh, tsk*, false starts) precede the initial clause of the resumption segment in 5 cases, and relatively long pauses (> 1.5 sec.) in 6 (cf. (Hirschberg and Grosz 1992). This suggests that in spontaneous oral discourse, instead of giving explicit indicators of the structural and semantic relations among segments, speakers provide non-lexical and pausal cues to breaks in segmental structure, relying on the hearer to infer the abstract structural and semantic relations.

For example, the semantics of the predication *fall over* at the onset of the resumption segment is arguably very dissimilar to the predications occurring in the intervening segment, possibly supporting the inference that the new clause is unrelated to the intervening segment. In contrast, the two clauses which end the suspended segment (at 14.1) and begin the resumption segment (at 16.1) are semantically identical and structurally parallel, supporting the inference that 16.1 resumes the segment containing 14.1.

In Section 5.1 we used the segment shown in Figure 6 to suggest that subjects agree on the narrator's intentions for reliable segments. In general, pops occur when there is particularly strong semantic discontinuity with the previous segment. Our data reflects this observation, in that the average strength of all boundaries separating intervening and resumption segments in our training corpus is 6.29, while the average strength of all empirically reliable boundaries is only 5.3. Furthermore, the intentions associated with intervening segments as a particularly similar across subjects, as illustrated previously in Figures 6 and 7, and as will be illustrated in the remainder of this section.

Consider the excerpt in Figure 9, where the narrator refers to a boy (who has fallen off his bicycle) in phrase 21.2, suspends this segment to describe a toy belonging to another boy, then resumes reference to the fallen boy in phrase 24.1 of the resumption segment. Note that the pronoun in 24.1 is not ambiguous between a reference to the first boy (mentioned in 21.2) and the second boy (mentioned more recently in 23.1 and 21.5), due to the "pop" of the intervening segment and the resumption of the suspended segment. The 5 subjects who agree on the intervening segment all indicate that the speaker's intentions pertain to

---

14 We assume that different uses of cue phrases can be discriminated; cf. (Hirschberg and Litman 1993).

describing the toy (while subject g further subdivided the segment, each of the 3 subintentions was about the toy). It is also interesting to note that 4 of the 5 subjects explicitly refer to the intervening nature of the segment (by using terms such as "suspends", "sidebar", "focus of attention is now", and "again").

21.1   [.4] Three boys came out,

21.2   [.75] helped him$_i$ pick himself up,

21.3   [.55] pick up his bike,

21.4   pick up the pears,

—5 SUBJECTS (b, c, d, e, g)—

21.5   [.55] one of them had [.6] a toy,

21.6   which was like a clapper.

—2 SUBJECTS (a, g)—

22.1   [2.4 [1.1] A-nd [1.1]] I don't know what you call it except
          a paddle with a ball suspended on a string.

—2 SUBJECTS (f, g)—

23.1   [.25 breath] So you could hear him playing with that.

—6 SUBJECTS (a, b, c, d, e, g)—

24.1   [2.25 [.8] A-nd [.95]] then he$_i$ rode off,

Subject	Annotation of Narrator's Intention for Intervening Segment
b	suspends description of events to describe objects boys had, and sound
c	sidebar description of a paddle ball
d	the focus of attention is now the toy
e	toy is introduced. Basically sensory aspects are mentioned again.
g	one of them had a toy; reader trying to name the toy; boy playing with toy

Figure 9. Intervening segment spanning 21.5 through 23.1

6. Conclusion

We have argued for the need for extensive, rigorous analyses of discourse corpora in order to determine the nature of the interaction of segmentation, coherence and linguistic devices in discourse. We have presented the results of a study of informal, spoken, monologic narratives that looks at certain aspects of this interaction. Our results show that naive subjects agree significantly often on segment boundaries, using an informal notion of speaker intention as the segmentation criterion. Given that subjects were allowed to use any number of segment boundaries in their linear segmentations, average IR scores across subjects at identifying boundaries will necessarily be far from ideal. Furthermore, there is considerable variation across narratives in the IR scores of individual subjects.

While none of our three algorithms—NP, CUE and PAUSE—performs as well as the human subjects, the results suggest that with additional knowledge, levels approaching human performance could be achieved algorithmically. However, we argue that no algorithm relying on a single class of linguistic device could hope to achieve high levels of performance across speakers due to variations in speaker strategy both across and within narratives. Another problem is that a discourse produced on one occasion may consist of distinct parts that are only weakly related to one another. Thus, when a coherent narrative episode is extracted from our corpus, the NP algorithm performs far better. Its performance improves even further when segment location is defined using fuzzy rather than precise boundary locations.

We believe that fuzzy boundary location is an important issue. Although the use of naive subjects necessarily produces noisy data, we believe that segment boundary location will turn out to be inherently imprecise. First, the role of certain utterances may be ambiguous, leading to indeterminacy regarding which segments they belong to. We illustrated one example (cf. Figure 7) in which a subject who identified a distinct initial segment boundary had a clearly divergent interpretation of the narrator's intentions. In this particular case, the subject's interpretation may reflect poor understanding, but in other cases, ambiguity may reflect equally good competing interpretations. Second, a single utterance may simultaneously have multiple functions, leading to distinct segmentations (cf. discussion of (Moore and Pollack 1992) in Section 2). Although natural language processing systems must produce a linear sequence of utterances as output, or produce an abstract model from a linear sequence of input utterances, the evidence suggests that ultimately it may be insufficient to assume that utterances contribute to or derive from only a single intention (or plan) at a time.

Speaker variation is another important issue with differing consequences for generation and understanding. Individual speakers have more or less skill at producing coherent discourse, and a similar point can be made regarding individual hearers. Understanding, generation or interactive systems will need to adapt these differences. Presumably, however, a generation system should produce discourse where choices regarding individual linguistic devices as well as linear sequence more closely resemble the performance of the 'best' or most easily understood speakers. Conversely, an understanding system should be designed to accommodate a wide range of hearers, not just the most adept.

Acknowledgments

This paper combines and expands on two previous papers: (Litman and Passonneau 1993) and (Passonneau and Litman 1993). The authors wish to thank W. Chafe, K. Church, J. DuBois, B. Gale, V. Hatzivassiloglou, M. Hearst, J. Hirschberg, J. Klavans, D. Lewis, E. Levy, K. McKeown, E. Siegel, and anonymous reviewers for helpful comments, references and resources. Both authors' work was partially supported by ARPA and ONR under contract N00014-89-J-1782; Passonneau was also partly supported by NSF grant IRI-91-13064.

References

Austin, J.L. 1962. *How to Do Things with Words*. Oxford and New York: Oxford University Press.

Beckman, M. 1991. Notes on prosody. Ms.

Butterworth, B. 1980. Evidence from pauses in speech. In B. Butterworth (ed.), *Language Production*. London: Academic Press. 155–176.

Carlson, G. 1977. A unified analysis of the English bare plural. *Linguistics and Philosophy* 1, 413–457.

Chafe, W.L. 1980. The deployment of consciousness in the production of a narrative. In W.L. Chafe (ed.), *The Pear Stories: Cognitive, Cultural and Linguistic Aspects of Narrative Production*. Norwood NJ: Ablex Publishing Corporation.

Chafe, W.L. 1980. *The Pear Stories: Cognitive, Cultural and Linguistic Aspects of Narrative Production*. Norwood NJ: Ablex Publishing Corporation.

Cochran, W.G. 1950. The comparison of percentages in matched samples. *Biometrika* 37, 256–266.

Cohen, R. 1984. A computational theory of the function of clue words in argument understanding. In *Proceedings of* COLING–84. Stanford CA. 251–258.

Fox, B.A. 1987. *Discourse Structure and Anaphora: Written and Conversational English*. Cambridge: Cambridge University Press.

Gale, W., K.W. Church and D. Yarowsky. 1992. Estimating upper and lower bounds on the performance of word-sense disambiguation programs. *In Proceedings of ACL*. Newark DE. 249–256.

Grosz, B.J. and J. Hirschberg. 1992. Some intonational characteristics of discourse structure. *In Proceedings of the International Conference on Spoken Language Processing*.

Grosz, B.J. 1977. *The Representation and Use of Focus in Dialogue Understanding*. Ph.D. dissertation, University of California, Berkeley.

Grosz, B.J. and C.L. Sidner. 1986. Attention, intentions and the structure of discourse. *Computational Linguistics* 12, 175–204.

Hawkins., J.A. 1978. *Definiteness and Indefiniteness*. NJ: Humanities Press.

Hearst, M.A. 1993. TextTiling: A quantitative approach to discourse segmentation. Technical Report 93/24, Sequoia 2000 Technical Report, University of California, Berkeley.

Hearst, M.A. 1994. Multi–paragraph segmentation of expository texts. Technical Report 94/790, Computer Science Division (EECS), University of California, Berkeley.

Heim, I. 1983. File change semantics and the familiarity theory of definiteness. In Bauerle, Schwarze, and von Stechow (eds.), *Meaning, Use and Interpretation of Language*. Berlin: Walter de Gruyter.

Hirschberg, J. and B.J. Grosz. 1992. Intonational features of local and global discourse structure. In *Proceedings of the DARPA Workshop on Speech and Natural Language*.

Hirschberg, J. and D. Litman. 1993. Empirical studies on the disambiguation of cue phrases. *Computational Linguistics* 19.

Hobbs, J.R. 1979. Coherence and coreference. *Cognitive Science* 3(1), 67–90.

Iwánska, L. 1993. Discourse structure in factual reporting (in prep.).

Johnson, N.S. 1985. Coding and analyzing experimental protocols. In T.A. Van Dijk (ed.), *Handbook of Discourse Analysis, Vol. 2: Dimensions of Discourse*. London: Academic Press.

Kamp, H. 1981. A theory of truth and semantic representation. In J. Groenendijk, T. Janssen, and M. Stokhof (eds.), *Formal Methods in the Study of Language, Part I*. Amsterdam: Mathematisch Centrum. 277–322.

Karttunen, L. 1976. Discourse referents. In J. McCawley (ed.), *Syntax and Semantics, Vol. 7: Notes from the Linguistic Underground*. New York: Academic Press,.

Krippendorff, K. 1980. *Content Analysis: An Introduction to Its Methodology*. Beverly Hills CA: Sage Publications.

Levy, E. 1984. *Communicating Thematic Structure in Narrative Discourse: The Use of Referring Terms and Gestures*. Ph.D. dissertation, University of Chicago.

Litman, D. and R. Passonneau. 1993. Empirical evidence for intention-based discourse segmentation. In *Proceedings of the ACL Workshop on Intentionality and Structure in Discourse Relations*.

Mann, W.C., C.M.I.M. Matthiessen, and S.A. Thompson. 1992. Rhetorical structure theory and text analysis. In W.C. Mann and S.A. Thompson (eds.), *Discourse Description*. Amsterdam: J. Benjamins Pub. Co.

Mann, W.C. and S.A. Thompson. 1988. Rhetorical structure theory: towards a functional theory of text organization. *Text* 243–281.

Moore, J.D. and C.L. Paris. 1993. Planning text for advisory dialogues: Capturing intentional and rhetorical information. *Computational Linguistics* 19, 652–694.

Moore, J.D. and M.E. Pollack. 1992. A problem for RST: The need for multi-level discourse analysis. *Computational Linguistics* 18, 537–544.

Morris, J. and G. Hirst. 1991. Lexical cohesion computed by thesaural relations as an indicator of the structure of text. *Computational Linguistics* 17, 21–48.

Passonneau, R.J. 1992. Getting and keeping the center of attention. In R. Weischedel and M. Bates (eds.), *Challenges in Natural Language Processing*. Cambridge: Cambridge University Press.

Passonneau, R.J. 1993. Coding scheme and algorithm for identification of discourse segment boundaries on the basis of the distribution of referential noun phrases. Technical report, Columbia University.

Passonneau, R.J. 1994. A plan based architecture for processing definite and indefinite descriptions in discourse. Manuscript in review.

Passonneau, R.J. and D. Litman. 1993. Intention-based segmentation: Reliability and correlation with linguistic cues. In *Proceedings of the 31st Annual Meeting of the. ACL*.

Pierrehumbert, J. 1980. *The Phonology and Phonetics of English Intonation*. Ph.D. dissertation, MIT.

Polanyi, L. 1988. A formal model of the structure of discourse. *Journal of Pragmatics* 12, 601–638.

Prince, E.F. 1981. Towards a taxonomy of given-new information. In P. Cole (ed.), *Radical Pragmatics*. New York: Academic Press. 223–255.

Reichman, R. 1985. *Getting Computers to Talk Like You and Me*. Cambrdige MA: MIT Press.

Rotondo, J.A. 1984. Clustering analysis of subject partitions of text. *Discourse Processes* 7, 69–88.

Schegloff, E. and H. Sacks. 1973. Opening up closings. *Semiotica* 8, 289–327.

(Passonneau) Schiffman, R.J. 1985. *Discourse Constraints on "it" and "that": A Study of Language Use in Career-Counseling Interviews*. Ph.D. dissertation, University of Chicago.

Sidner, C.L. 1979. Towards a computational theory of definite anaphora comprehension in English discourse. Technical report, MIT AI Laboratory.

Webber, B.L. 1987. A formal approach to discourse anaphora. Technical Report 3761, Bolt Beranek and Newman Inc.

Webber, B.L. 1991. Structure and ostension in the interpretation of discourse deixis. *Language and Cognitive Processes*, 107–135.

Authors' Biographies

Kathleen Dahlgren holds a Ph.D. in Linguistics from the University of California in Los Angeles (UCLA) in semantics. Her research has concentrated on Computational Linguistics, formal semantics, and Psycholinguistics. After completing the Ph.D. in 1976, she taught Computer Science at California State University at Northridge, and Linguistics, including Computational Linguistics, at Pitzer College of the Claremont Colleges in Los Angeles. She completed a postdoctoral degree in Computer Science at UCLA in 1980. From 1984 to 1990, she led a natural language understanding research project at the IBM Los Angeles Scientific Center. Presently, Dr. Dahlgren serves as President and co-founder of Intelligent Text Processing, Inc., a start-up company with a text retrieval product called InQuizit that employs full language understanding techniques. Dr. Dahlgren is the author of a book on computational semantics (Naive Semantics for Natural Language) and of articles and papers on semantics and Computational Linguistics.

Dr. Kathleen Dahlgren
Intelligent Text Processing Inc.
1310 Montana Avenua, Suite 201
Santa Monica, CA 90403
U.S.A.
email: kd@itpinc.com

Eva Hajicová is a member of the Institute of Formal and Applied Linguistics at Charles University in Prague. One of the principal figures of the Prague School, she has contributed especially to the understanding of Topic, Focus, Given, and New, publishing in the fields of Linguistics, Computational Linguistics, and Artificial Intelligence. A paper describing an automatic procedure to identify topic and focus elements recently appeared in the journal *Computational Linguistics*.

Prof. Eva Hajicová
Institute of Formal and Applied Linguistics
Faculty of Mathematics and Physics
Charles University
Malostranské námesti 25
CS-118 00 Prague 1
Czech Republic
email: hajicova@cspguk11.bitnet

Jerry Hobbs is the co-author and co-editor of several books, including *Formal Theories of the Commonsense World* (with R.C. Moore), as well as of numerous journal and book articles. Since completing a Ph.D. at City College of New York in the late 1970s and teaching at Yale University, Dr. Hobbs has been a member of the research staff at SRI International, a well-known research laboratory near San Francisco. He has published in areas as diverse as semantics, discourse, parsing, and knowledge representation, and is currently actively developing the notion of semantic analysis as abductive inference. Dr. Hobbs is a Fellow of the Association of Artificial Intelligence and has served as President of the Association of Computational Linguistics.

Dr. Jerry Hobbs
SRI International
333 Ravenswood Avenue
Menlo Park, CA 94025
U.S.A.
email: hobbs@ai.sri.com

Diane Litman is a Member of Technical Staff in the Artificial Intelligence Principles Research Department of AT&T Bell Laboratories. She received Ph.D. and M.S. degrees in Computer Science from the University of Rochester in 1986 and 1982, and a B.A. in Mathematics and in Computer Science from the College of William and Mary in 1980. Dr. Litman's research spans several areas of Artificial Intelligence, including spoken and written natural language processing, knowledge representation and reasoning, and plan recognition. Dr. Litman is currently the book review editor of the journal *User Modeling and User Adapted Interaction*, has served on the editorial board of the journal *Computational Linguistics*, and has both chaired and been a member of numerous program committees. From 1990 to 1992, Dr. Litman was also an Assistant Professor of Computer Science at Columbia University, New York.

Dr. Diane Litman
AT&T Bell Laboratories
600 Mountain Ave
Murray Hill, NJ 07974
U.S.A.
email: diane@research.att.com

Jim Martin is currently Associate Professor of Linguistics at the University of Sydney, Australia. His research interests include theory in Systemic Functional Grammar, functional grammar, discourse analysis, register, genre, and ideology, focussing on English and Tagalog, with special reference to the transdisciplinary fields of educational linguistics and social semiotics. Recent publications include the following books: the programme *Language: A Resource for Meaning* from Harcourt Brace Jovanovich (co-authored with Frances Christie, Brian Gray, Pam Gray, Mary Macken, and Joan Rothery) 1990, 1992; *English Text: System and Structure* from Benjamins, 1992; *Writing Science: literacy and discursive power* (with M.A.K. Halliday) from Falmer, 1993; *Deploying Functional Grammer* (with C.M.I.M. Matthiessen and C. Painter) from Edward Arnold Press, in press; *Genres and Institutions: Social Processes in the Workplace and School* (co-edited with F. Christie) from Pinter Press, in press.

Prof. James Martin
Department of Linguistics
University of Sydney
Sydney, NSW
Australia
email: jmartin@extro.ucc.su.oz.au

Tsuyoshi Ono is a faculty memeber at the University of Arizona where he teaches Japanese language and linguistics. He holds a Ph.D. in Linguistics from the University of California in Santa Barbara; his dissertation was on the discourse and grammar of Barbaren~o Chumash (a California language). He has worked extensively on grammar and interaction in English and Japanese, and he is currently building a corpus of Japanese conversation. His research interests also include language change and research methodology, and in conjunction with Prof. Sandra Thompson he is working on a model of syntax which is adequate to account for conversational data.

Prof. Tsuyoshi Ono
Department of East Asian Studies
Franklin Hall 404
University of Arizona
Tucson, AZ 85721
U.S.A.
email: ono@u.arizona.edu

Rebecca Passonneau is currently a Resident Visitor in the Information Sciences Research Group at Bellcore and a Research Scientist in the Department of Computer Science at Columbia University. From 1985 through 1990, she worked in the Natural Language Group at the Paoli Research Center, Unisys. From the University of Chicago she received a Ph.D. (1985), M.A. (1976), and B.A. (1974) degrees in Linguistics. Dr. Passonneau's research interests include discourse modelling, processing of nominal and temporal reference, lexical semantics, and empirical methods for analysis of discourse corpora. She has served on program committees for AAAI, ACL, and has served on the editorial board of *Computational Linguistic.*

Dr. Rebecca Passonneau
Department of Computer Science
Columbia University
New York, NY 10027
U.S.A.
email: becky@cs.columbia.edu

Emanuel Schegloff is Professor of Sociology and Co-Director of the Center for Language, Interaction and Culture at the University of California, Los Angeles. He received his B.A. degree with honors from Harvard in 1958 and M.A. (1960) and Ph.D. (1967) degrees in Sociology from the University of California, Berkeley. Before moving to UCLA, he taught at Columbia University in the City of New York. The author or co-author of some 50 papers, Dr. Schegloff is co-editor (with Elinor Ochs and Sandra Thompson) of the volume Interaction and Grammar (Cambridge University Press). Dr. Schegloff's research interests—focused on forms, settings, and structures of talk-in-interaction—span aspects of Anthropology, Applied Linguistics, Communications, Linguistics, Psychology, and Sociology. He serves on advisory and editorial committees in several of these fields. He has been a Fellow of the Netherlands Institute for Advanced Study in the Social Sciences and Humanities, Starr Lecturer in Linguistics at Middlebury College, McGovern Lecturer in Communications at the University of Texas

(Austin), and Brittingham Lecturer in Sociology at the University of Wisconsin Madison, and is a member of the Sociological Research Association.

Prof. Emanuel Schegloff
Department of Sociology, 264 Haines Hall
University of California in Los Angeles
Los Angeles, CA 90095-1551
U.S.A.
email: scheglof@soc.sscnet.ucla.edu

Sandra Thompson is Professor of Linguistics at the University of California at Santa Barbara. She received BA, MA, and Ph.D. degrees in Linguistics from Ohio State University. She has authored, co-authored, and co-edited numerous articles and books, and she was the recipient of a Guggenheim Fellowship in 1989 for her research on language universals. Dr. Thompson specializes in Chinese linguistics, discourse, and language universals, particularly on the role of patterns of conversational discourse in shaping morphosyntactic regularities. She is the co-author with Charles Li of *Mandarin Chinese: A Functional Reference Grammar*. She has co-edited *Studies in Transitivity* with Paul Hopper, *Clause Combining in Grammar and Discourse* with John Haiman, *Discourse Description* with William C. Mann, and *Interaction and Grammar* with Elinor Ochs and Emanuel Schegloff. Prof. Thompson has also published numerous papers in journals and books.

Prof. Sandra Thompson
Department of Linguistics
South Hall
University of California in Santa Barbara
Santa Barbara, CA 93106
U.S.A.
email: sathomps@humanitas.ucsb.edu

Index